You and Your Big Ideas!

The Handy Pocket Resource Guide

For

Inventors, Innovators & Entrepreneurs

Invented By Brian Fried

© 2008

Copyright © 2008, Brian Fried. All rights reserved.

Parts of this book up to 100 words may be quoted freely without further permission of the author by providing the name of this book and its author for full credit. Beyond above permission for limited reproduction, all rights reserved by contacting author directly at Contact: legal@youandyourbigideas.com

Printed in the United States of America
Published by WingSpan Press, Livermore, CA
www.wingspanpress.com
The WingSpan name, logo and colophon are the trademarks of WingSpan Publishing.

Special thank you to Lisa Mirabile of Vertigo New York for the outstanding book cover design.
www.vertigonewyork.com

First Edition 2008

ISBN 978-1-59594-260-9

Library of Congress Control Number: 2008936865

The following opinions are based on personal experiences and are being shared with you to make educated decisions during your path to inventing. It is sold with the understanding that the publisher is not engaged in rendering legal, accounting, or other professional advice. If legal advice or other expert assistance is required, the services of a competent professional person should be sought. You are ultimately responsible for your own actions.

Contact the author at brian@youandyourbigideas.com

This book is available at volume discount for bulk purchases.
Contact: **orders@youandyourbigideas.com** or
call 888-64-Think (888-648-4465)

From the desk of Brian Fried

I love coming up with a fresh idea or a better way to overcome a challenge! After the AHA moment, I do my due diligence. I give it a name. I share it with my wife and daughter and get feedback from my circle. Whether they say, 'Hey, that's great!' or 'Are you nuts?' it means I get to move on to the next step.

What makes me crazy is when I don't know what to do next; when I hit a wall because I can't figure out that next step... which can happen whether you're just getting started or you've been around the block a time or two.

I'm not talking about normal delays. Step-by-step processes take time and it's often the achievement of those steps... when we get patent approvals; when see the engineer's drawings; when the prototype is in our hands; when our big idea is finally on the shelf... this is when success tastes great!

I'm not talking about the uncertainties that might stop you from taking the plunge. If you're having one now, this is your golden opportunity to discover how easy it is to pass this test with flying colors and move on to the next, living a life as exciting and exhilarating as a roller coaster ride that you wouldn't change for anything in the world.

What made me put my big idea-ing on hold to write this book is the frustration I feel when the only thing standing between me and my next step is having no idea where to go, who to turn to or what to do. It grew out of conversations with other big idea people who said, 'It makes me crazy too.' That's when I decided that this book was long overdue.

This book is for those who get that special feeling when you know you're onto something good; who can't wait to hear people ask 'Why?' and 'How?' because it's a chance to share the magic; who live for the moment when you make your families proud; who know that a 'WOW!' from total strangers is the best feeling in the whole world!

This book was invented to help you overcome the 'Now what' moments that can slow you down and help you build your networks to get the answers you need to reach your goals.

Good luck on your journey!

Acknowledgments

I would like to give special thanks to my wife Lisa and my little inventor Alana, who have and continue to put up with me and my big ideas! Both of you have made many sacrifices along the way in order for me to achieve levels of success and have been there to support me along the way. I love my girls.

I would also like to acknowledge my Parents Alex and Penina, for always being there for me no matter what, while exploring the path of life and giving me that opportunity. Thank you for being such great parents, I love you both. Parent-in- Laws Janet and Kenny, brother and sister Robert and Jennifer, thank you for always giving me an open ear for me to run ideas past them no matter how great or crazy they are. All of you have been there to help me and I appreciate your honest opinions whenever I ask, any time of the day, late night and even at 3am!

To my Safta, your lifelong personal struggles and facing them have been inspirational to me in more ways than you can imagine. You, Saba, Bobbi and Zayde (who I know always watch over me) have faced and overcome obstacles beyond imagination. It is with your courage instilled in me that I continue to strive for successes without fear of roadblocks that may appear. I know that I am making you all proud of me.

A special thank you to my Brother-In-Law Michael, who has and always will be a major influence in my life personally and professionally.

I would also like show my appreciation to Abby Hagyard, for her hard work and dedication while working with me to make sure this book reached fruition above my highest expectations!

Thank you to all my inventor friends I have met and continue to meet along the way. Let's all continue our camaraderie and brainstorming to help each other and benefit others. Let's always be creative and inventive so we can have stories of success to share through our continued friendship. Now let's get started with You & Your Big Ideas....

TABLE OF CONTENTS

INTRODUCTION v
PART ONE: PREPARATION 1
1. Are you an Inventor? 3
2. Rate Your Attitude 7
3. Find a Mentor 11
4. Educate Yourself for Success 15

PART TWO: GETTING STARTED 21
5. You Have an Idea! Now What? 22
6. Protecting Yourself 29
7. Making It Real 37
8. Making It Work 40

PART THREE: THE FORK IN THE ROAD 47
9. Taking Care of Business 48
10. If You Decide to Manufacture 58
11. If You Decide to License 64
12. How to Work as a Team 66

PART FOUR: THE ENTREPRENEUR 69
13. Farming 71
14. Fishing 74
15. Hunting 77
16. Nurturing 84

PART FIVE: LIVING THE DREAM 87
17. Packaging 88
18. Promoting 92
19. Pitching 95
20. The Perpetual Growth Machine 97

CONCLUSION 99
BEST BETS 102
RESOURCE GUIDE 127

INTRODUCTION

What does it take to be a successful inventor?

Your first thought is probably going to be 'great ideas' and you're right. Having an inventive mind is essential. A positive attitude, a willingness to work hard and a determination to reach your goal are key ingredients as well. But I believe there is something else that every inventor needs if he or she hopes to achieve success. I believe it takes teamwork – the kind of teamwork you find in a strong support network of colleagues and collateral services.

As an innovative thinker, you probably do most of your creative thinking alone. You seldom have a chance to enjoy the benefits that a team of supporters can bring to the mix. You miss out on the legacy of learning from others who have stood where you are standing now. You don't have the opportunity to brainstorm ideas and leverage information to fast-track your journey to success.

This book is designed to change all that. Each page is filled with insider tips and 'been there, done that' advice that will help you through the awkward beginner phase and set you firmly on the path to success.

A little background on me

I'm an inventor from Suffolk County, New York. I live in a community where brilliant innovations in medical science, aviation and the aerospace industry first got their start. Under the circumstances, it would be logical to assume that I would have all the tools of the trade at my fingertips. But that's not the case. When I began to work toward my dream of becoming an inventor five years ago, I had no idea what to do or where to start. I made my fair share of mistakes along the way and learned a lot of valuable lessons as well.

The valuable lessons I learned enabled me to achieve some pretty amazing results in a relatively short period of time. I've patented three products and registered 150 trademarks; I've also got six other products patent-pending.

Doing the research and learning from my mistakes ate up a lot of valuable time because I didn't have access to the material I've pulled together here. I didn't have a virtual team contributing their information and ideas to help me make the decisions that were right for me. I didn't have a support group that had my best interests at heart.

If you've already started doing your homework to take your 'big idea' to the next level, you know there are many challenges to overcome. Sometimes it's tough to keep a positive attitude and stay focused on your dream.

Being an inventive thinker, I suddenly realized that if I wanted a support group to help me keep that positive attitude and drive that forward momentum, I would have to create it myself... so I did! In 2007, I encouraged Suffolk County to sponsor a not-for-profit Inventors and Entrepreneurs Club that provides a place where inventors can brainstorm with others who share similar interests or goals. They can learn about services and educate themselves about the various legal processes that must be followed. They can avoid costly mistakes and steer themselves clear of 'sharp practices' and unscrupulous individuals that take advantage of beginners.

You can visit the Inventors and Entrepreneurs Club of Suffolk online by going to the website: **www.iesuffolk.com** or by emailing: **info@iesuffolk.com**.

The next logical step

You know how inventors are. They get one big idea and before you know it, they get another one. It was while I was presenting a lecture series that highlights the essentials an inventor needs to know that the idea of a *portable* support network came to me. For those inventors and entrepreneurs who couldn't access our group or come to my lectures when they needed answers, a hands-on mentoring guide would be an extremely valuable resource. And that's how I came to write this book.

Inventing a new book is like inventing anything… once you get your big idea, the first thing you need to do is make sure you're offering a unique, stand-alone solution no one else has thought of in quite the same way. So I did my due diligence and explored the book shelves. What I discovered were a lot of great books designed to inform and educate the inventor, but very few that give an insider's point of view and almost none that handed you the tools you need, when you need them, to take the next step.

Because I've been there before and I'll be there again, I know exactly what you're going through. I know the questions you're asking and the frustration you're feeling when the answers just don't provide the practical knowledge that you are desperately looking for NOW.

You and Your Big Ideas! is all about you. Each chapter contains valuable information and advice that will allow you to achieve your goals sooner rather than later and at much less cost (emotional and financial) to you. To help you make your way easily and quickly through the contents, the book is divided into five parts. It also has a Best Bets section where I introduce the go-to people who have consistently been there for me and helped me reach my dreams while they simplified my life. There is a handy Resource Guide located at the back where you can quickly access the contact information mentioned throughout the book.

Part One is designed as a Primer/Refresher tool. Each chapter highlights the qualities and practices that successful Inventors share. You may not feel that *you* need this information but it won't hurt to take a look. If you already have a firm grasp on this material, it's quite likely that your friends and family will benefit from seeing that your methods and mindset aren't proof of madness after all; that they are the necessary ingredients in the recipe for an inventor's success. You need their support. You need their positive energy. You need to have them cheering from the sidelines when the going gets tough.

You need to have them understand why you are driven to do what you do and stop saying things like, *"Why are you still playing with that crazy idea?"* and *"What makes you think anyone will buy it?"* and *"When are you coming to bed?"*

Part Two contains four chapters that provide essential information to help you master those all-important first steps. You would be amazed at how many people get so carried away with the excitement of an idea that they forget the 'walk before you run' rule. Their big idea is so real to them that they gloss over the details in their eagerness to reach their goal. These are the people whose ideas get stolen; whose energies get compromised and whose dreams get broken. They will tell you that they have been victimized when more often than not, it turns out that their bad luck is the result of their own reluctance to walk before they run.

In **Part Three** we take you through the development process. We've called this section 'The Fork in the Road' because this is where you have to make sensible choices – Are you qualified to do it yourself? Is it time to call for reinforcements? How will you structure the give-and-take process? When you plan to launch something ambitious that requires a significant investment of time, money and talent, you need to accept the fact that no one can do everything himself. You need to be willing to accept the fact that your inexperience and under developed skill sets can get in the way of your success.

Part Four takes you through a difficult transition – from Innovative Thinker to Entrepreneurial Do-er – providing the essential tools that every entrepreneur needs in order to achieve success. It's not enough to know *what* needs to be done. The successful entrepreneur also has to know *how, why* and *when*. He has to know *who* the best person is for the job.

Part Five is your guide to long-term growth and bottom-line success. All too often the novice inventor believes that when the invention is out there, the job is done. The chapters in this section will help you take control of your future by helping you manage your business with maximum efficiency. This section takes you out of the realm of inventing and into the world of building and maintaining the perpetual growth machine.

Our **Best Bets and Resource Guide** is provided to save you time and trouble. When you're stuck and you need the name of that special someone who can leverage your success... when you need facts, figures and contact information, you need them NOW. You don't have time to flip through the entire book, so we've organized everything in one place to make life simpler for you.

And finally...

From where I sit, being an Inventor is the greatest gift and the craziest curse in the world. I have more fun than anyone I know. I have more frustration as well.

The pride I feel when a big idea of mine has taken on a life of its own is unlike anything I've ever known. Knowing that it's happened before and can happen again is the best adrenalin rush there is.

Being an Inventor has changed my life. What started out as a hobby for me has become a thriving business: Today, ThinkUp Designs [www.thinkupdesigns.com] keeps me busy round the clock. Just imagine what your big ideas will do for you!

Brian Fried

PART ONE: PREPARATION

How many times have you been told to take it easy... take it slow... to walk before you run... and don't put the cart before the horse?

How many times did you grit your teeth and swallow your impatient reply?

That's good. That means you've learned to be patient with the people who have your best interests at heart.

With any luck, it also means that you understand the dangers that lie ahead for those who rush into things blindly without taking the time to prepare for the journey that lies ahead.

The four chapters included in this section are designed to prepare you for the best, most challenging adventure of your life. We want you to enjoy it to the fullest and have the most fun you can. So take a deep breath, put on that patient smile you've perfected and let's get you started on the road to success...

CHAPTER ONE: ARE YOU AN INVENTOR?

THINK ABOUT IT...

If you've picked up this book you already know the answer, but as any successful inventor will tell you... the research you do *before* you set out on your journey is the step that will bring you closest to your goal. So let's take a moment and ask...

- **Do you have an inventive mind?**
    - Do you come up with ideas and solutions to challenges on a daily basis?
        - Inventive minds are always seeing the next logical step that could be taken, while the rest of the world simply shrugs its shoulders and makes do with things as they are.
    - Do you say 'This could be done differently...'; 'That could be improved...'; 'This could be modified a bit...' or 'That could be a big winner'?
        - Inventive minds get excited when there is a puzzle to solve.
    - Do you see opportunity everywhere you look?
        - Inventive minds are positive and pro-active. They consistently ask themselves, 'Why not?' when other people give up.
- **Do you have inventive habits?**

You And Your Big Ideas

- Are you curious?
    - ☐ Are you the kind of person who asks questions?
- Do you find yourself wondering how things work?
    - ☐ Do you take things apart and see if they'll go back together in a different, more efficient way?
- Do you make notes and sketches and conduct experiments?
    - ☐ Do you like to keep track of your ideas and use your notes and sketches to fix and tweak them from time to time?
- Do you brainstorm with others and get them excited by your ideas?
    - ☐ Is it as much fun to talk about the process as it is to work on it?
    - ☐ Do you find yourself feeling energized when you share your ideas and get input from others?
- Are working on some idea that you're nearly ready to talk about?
    - ☐ Have you done your due diligence?
    - ☐ Have you researched the market and compared your concept with others currently on the shelves?
    - ☐ Have you registered your patent to protect your concept?
- Are you at the prototyping stage?
    - ☐ Have you created a sample that allows other people to see and touch your idea and believe in its possibilities as you do?
    - ☐ Do you have a design engineer who is willing and able to take your prototype and create a professional design?
    - ☐ Have you looked for opportunities to partner with

a designer and take advantage of his established contacts?

- Are you ready to go to the marketplace?
    - ☐ Have you decided whether you want to manufacture it yourself or license it to someone else and let them pay you royalties?
    - ☐ Have you registered your business and surrounded yourself with the best team to help you reach your goals?

☐ ***Do you have a goal?***

- Do you want to solve problems and overcome challenges?
    - ☐ Are you driven by a desire to make the world a better place?
    - ☐ Are you searching for opportunities to beat the odds and win?
- Do you want the personal satisfaction of a job well done?
    - ☐ Are you the kind of person who sets high goals and standards and consistently meets and surpasses them?
- Do you want to be famous?
    - ☐ Are you looking for the recognition that comes with success?
- Do you want to be rich?
    - ☐ Are you focused on the tangible rewards that come with high volume sales and repeat business?

☐ ***Do you have what it takes to achieve success?***

To take a great idea or concept from brain wave to prototype to marketplace, it takes more than dedication, determination and drive. To do your due diligence, to research and refine, to educate and inform, to overcome setbacks, to protect your work, to form partnerships and leverage your success… it takes more than perseverance and people skills.

Whether you're inventive and imaginative; whether you brainstorm or work alone; whether you're still thinking about taking that first step or you're already working on your big idea, the one quality you have to have in order to succeed as an inventor is patience!

What does that mean in tangible terms? You need to be able to count to ten. You need to be willing to walk away when that little voice tells you something's not right; when that feeling in your gut tells you this person, this process, this deal is essentially flawed. You need to be able to take a deep breath, clear your head and start again. You need to be able to laugh at yourself. You need to feel the joy of being an inventor instead of the fear and stress that opportunity is passing you by. You need to recognize and remember that this gift will always be there… if it's not this big idea, it will be the next one that delivers results.

Where does patience come from? It comes from believing that a thing worth having is worth working for. It comes from wanting to be the best instead of the first. It comes from believing in yourself – knowing that you have greatness within you. It takes a good attitude.

CHAPTER TWO: RATE YOUR ATTITUDE

How many times have you heard someone say, "Attitude is Everything"?

In fact, they're wrong. Timing is everything. Getting the right idea at the right time... finding the right people to supply the right support services... seeing the right application and carving the perfect market niche... this is everything.

But you can't do any of those 'right' things with the wrong attitude. You can't infect people with your enthusiasm if you're full of doubt... you can't inspire people to believe if you have no confidence in yourself... you can't free your mind to create timely solutions if you're desperately watching the clock.

Like 'timing', 'attitude' is a quality that is difficult to define because it presents itself differently in each of us. You may be the quietly enthusiastic type... you might be a ball of energy... you might be a loner who craves solitude or a nurturing team leader who generates crowds of followers everywhere you go.

Each one of these personalities has the 'right' attitude if it works for them.

So the question we want to address in this chapter is, 'How can you find and sustain the attitude that works for you'? The answer has to do with managing the ebb and flow of emotions that will surround you throughout the Inventing process.

As you proceed from that first 'AHA!' moment through all the steps and stages that follow, you will meet every personality there is. You will come face to face with doubters and doom-sayers... you will endure obstructive nit-pickers... you will be befriended by the untrustworthy and abandoned by your friends... and every now and then you will come across someone bright and forward-thinking who listens and looks and says, "What a great idea... how can I help?"

Inventors don't live in isolation from the world. They are fathers and mothers; sons and daughters; sisters and brothers; friends and colleagues with lives that cannot be put on hold. You cannot commit 100% of your time and energy to deal with the ups and downs of being an inventor because you've made commitments that must come first. You have a job to go to, bills to pay and people who have invested themselves in you. Even on a good day there is a certain amount of stress that comes with living. Even on a great day there are setbacks and disappointments that can undermine your positive attitude.

How do you manage all this conflicting input? How do you balance the ups and the downs and ensure that the bad never outweighs the good?

Success is not an overnight thing. For every step forward, the brightest minds and greatest inventors have taken a thousand steps back. It would take Thomas Edison more than 10,000 tries to perfect his light bulb invention. When he was asked how it felt to fail 10,000 times he replied, "I have not failed. I have succeeded in finding 10,000 ways that won't work."

Persistent vision and positive outlook do not happen by themselves. They are driven by behaviors that are purposefully installed into your daily life. Studies have shown that a lifetime habit can be broken in as little as three weeks simply by replacing the old habit with a new one. To help you learn and apply positive persistence to your life, your work and your big ideas, we offer you this proven 21-day five-step program... you will be amazed at the changes you will see in yourself after just three weeks by adopting these positive habits!

Step 1: Take charge of what you're thinking.

It's your brain. You get to decide what you think. What you think determines how you feel. Instead of letting outside actions and events control your mood and undermine your confidence, make the conscious decision to look for opportunity in every obstacle that comes along. Choose to distance yourself from negative thinkers and unproductive conversations. You need to make the conscious decision to walk away when everyone wants to gossip and criticize. Concentrate on actions that generate change instead of unhelpful complaints that keep you rooted in failure. Start and end each day with a positive statement. Look in the mirror and tell yourself one positive thing first thing in the morning and last thing at night. Keep a journal of every positive thing that happens each day. Make a habit of reviewing it every week and count your blessings that so many good things have filled your days.

Step 2: Find and share positive ideas, information and entertainment.

Get out of the habit of catching up on bad news first thing in the morning. It will still be there after your morning pick-me-up. Seek out books, music, talk shows, speakers or Internet sites that inspire you to feel good about yourself and your life. Whether you are religious or not, there is material that can inspire everyone. Increase your opportunities to smile and laugh. Listen to up-tempo or soothing music. Encourage your family and friends to contribute positive comments, inspiring stories or funny moments to conversations they have with you each day.

Step 3: Focus on kindness.

Make an extra effort to help others. A selfless act that enhances someone else's life does wonders for your self-esteem. Hold a door open, pull out a chair, pick up after someone else, offer a compliment, purchase an inexpensive treat or gift, thank someone for a small kindness, smile at every opportunity and make direct eye contact. Concentrating on other people allows you to step outside yourself and get a better perspective on the things that are troubling you.

Step 4: Take care of your health.

Kindness has to begin with you. It's hard to stay positive when you don't feel well and neglecting your basic needs while you serve others is a recipe for failure if ever there was one. While you're picking up that inexpensive treat for someone else, treat yourself. Eat sensibly. Get enough sleep. Take a few minutes out of your day to stretch and exercise some part of that amazing machine that works for you around the clock. Take a walk, close your eyes and take a few deep breaths, drink plenty of water… and smile! Smiling activates positive receptors that make everyone feel better… about themselves and you!

Step 5: Learn to be thankful.

Focus on what you have. Live in the present and enjoy your blessings. Now is the time to forget about keeping up with other people and be thankful for all that you have that makes your life truly special. Even your most challenging uphill battles can turn out to be valuable gifts when you look back on the lessons you learned as you met and mastered those hurdles. These lessons will enable you to help others when they need you most.

Taken together, these five steps will help you acquire the positive attitude habit that will change your life. Apply them every day for 21 days. Keep a journal of your progress and congratulate yourself when you feel great.

It will take patience and practice, but you're worth it!

Now is the time to choose your start date. Will it be today?

CHAPTER THREE: FIND A MENTOR

A mentor is someone who has walked a mile in your shoes. Because he or she has stood where you are standing now, the wisdom you can acquire from a mentor is not random knowledge – it is specific to your needs.

A mentor is a sounding board. You can share your ideas in their formative stages and gain valuable insights and fresh perspectives from someone who knows where you are coming from and where you hope to go.

A mentor holds the key to your future networks of colleagues and collaborators. His or her stature in the community will leverage your credibility; introductions he or she makes will open new doors for you.

And finally, a mentor is a friend… someone who has your best interests at heart; someone whose agenda includes you.

Where can you find a mentor? You will come across likely candidates each time you speak to an established professional whose work and work ethic impresses you. How do you get them to help you? You ask. You'll be surprised at how eager many of them will be to invest their time and energy in someone who is working hard to achieve a dream. Why will they do that? Because someone once invested their time and energy in them.

It may sound a little crazy, but the best way to find the mentor who will be the most helpful to you is to think of the people who have the most impact on you when you're all alone, reading and researching. I'm talking about the authors whose books and articles and opinions make the most sense to you. If their *words* have that kind of influence on you, imagine how helpful it would be if you could pick up the phone and get their insights and advice when a particular issue completely baffles you!

So how do you make that connection? In many cases, they're in business themselves. Their contact information is readily available, either because it is attached to their book, their article or their blog. You can also contact the source that published them. Newspapers and periodicals, information websites, TV and radio shows all have public access information. Publishers usually have an "About the Author" page on their websites as well.

Before you pick up the phone or send that email inquiry, however, make sure you know what you'd like this individual to do for you. Many 'experts' offer their services for an hourly fee, so don't be surprised if your query generates a price list of consulting services in return. Others may have agendas and are using their books and advice columns to build ancillary businesses. This may not be the kind of relationship you are looking for. Still others may be willing to work with you in exchange for a percentage of the royalties your product will earn; they may even be interested in a percentage of your business, if you have more than one concept in development and your company is well grounded.

Another excellent source for mentors is the list of people you are calling to get information regarding design engineering, licensing, manufacturing and distribution. Every now and then you will be speaking to someone you feel a real connection with and you'll hear that little voice inside your head saying, 'I wish this was my mentor!' When that happens, the first thing you should do is ask them if they ever take on the role of mentor and if so, would they be interested in mentoring you. In exchange for their help, you can offer them percentages on your royalties and/or in your business.

A mentor is someone who can guide you through both the business management *and* the entrepreneurial side of things.

Business coaches make a living doing this. You need to do some research, make some queries and select the one who understands the kind of person you are and whose strategies are compatible with your needs.

This last point is very important. You need a business coach with a strategic plan that is a good fit for you. Someone who has an entirely different attack than you do is not going to help you succeed. If you are low key and he or she sells a high energy, high pressure approach, it's not going to work. If you're upbeat and outgoing and your coach sells a low key management model, you'll get the same less than perfect fit. You will end up feeling frustrated with techniques you don't like. Time and money will be wasted and your business will suffer as a result.

Another very important point is that you can have more than one mentor. When you're starting out in business, success is much easier to achieve when you create a strong support network – a powerful team that is focused on helping you reach your goal. This is why I like to think of the mentoring process as a two-way street: Everyone is working together as a team that is focused on a goal that benefits each of them, for their own reasons and in their own separate ways. The people you hire add you to their list of satisfied clients. Your mentors exchange their information for a return that enhances their portfolio in more ways than one. As the catalyst that brings them together, you leverage the knowledge and experience of experts who have walked a mile in your shoes to reach your goal.

When the time comes to contact potential mentors, simple, clearly worded query letters provide the information necessary to help the reader say Yes to your request. Take the time to get it right... remember that you only get one chance to make a great first impression. Sample query letter to potential mentor:

You And Your Big Ideas

Hi Mr. _____

I've taken the liberty of contacting you because I understand you're very experienced in ... .

[Explain how you know this... you read their book, saw their interview, spoke to them at a trade show, called them for information, etc.]

I have a particular product I am looking to bring to market and I think it has a lot of potential. I would value any input, advice or support you might be able to share. In exchange for your contribution as a mentor, I offer a percentage of royalties once it's in manufacturing or licensing. I am also willing to offer a percentage of my company because I have several products in various stages of development that would benefit from your input.

Thank you for taking the time.

Sincerely,

_____

[Your contact information here.]

CHAPTER FOUR:

EDUCATE YOURSELF FOR SUCCESS

We've talked earlier about people who allowed their impatience and excitement to run away with their brilliant idea, landing themselves in unhappy and difficult situations that could easily have been avoided if they had taken the time to perform all the suggested necessary steps in the inventing process.

In this chapter we're going to talk about grounding yourself and your brilliant idea in reality, doing what is called "due diligence" and educating yourself for success.

The first thing that needs to be addressed is probably the most important one of all. Next to the brilliant idea itself, the one absolute essential you must familiarize yourself with is the bottom line. I would be doing you a disservice if I didn't help you get a solid grasp of the kind of money and time it can traditionally take from first AHA! to seeing your invention being sold.

I believe this is a critical step because an inventor invests more than time and money in his inventions: He invests a part of himself as well. When he or she under-estimates the costs involved – in both time and money – and the project comes to an abrupt halt before the dream has been realized, the disappointment can be devastating. Some people never stop blaming themselves for failing to ask the right questions and prepare themselves for the challenges to come.

As your unofficial mentor, I see it as my job to point out these challenges and offer my best advice to help you make the choices that will increase your chances of reaching your goal. I've made it a point to make sure every section and chapter of this book addresses this task because you are not expected to automatically know what it takes to be an inventor. There's no way you could know how much the various processes and procedures cost and the fees people charge to provide the necessary services that must be done.

Until I found myself standing where you are now, I had no idea what anything cost or how long it would take. I was completely in the dark when it came to getting the legal supporting work done... hiring experts in any number of fields... how to find these experts... how to get multiple quotes... how to validate their bona fides... how long it can take for a design to be done... how many you get for the price... how to negotiate that deal... how to draw up a letter of understanding for the services the expert will provide... how to build penalties into the agreement if the work they deliver is flawed... how to set up a business... how to run a business... and so on.

You can't be expected to know how much it can cost to create a rough sample or mock up of an idea. Each invention is unique in the materials it requires and the number of steps it takes. It's impossible to tell how many times you may have to revisit your local Target or Walmart, for example, until you get it right. Most people have their vision so firmly fixed in their minds, they don't realize how many steps are involved in actually building a prototype until they've started to build it themselves. If you don't build a margin of error into your cost estimate for your prototype, for example, and you're working with a limited budget, you're going to be in trouble almost before you start.

Not everyone will be willing to step up to the plate for a percentage either. All the positive encouragement I give people about mentors and their willingness to get involved in your dream is absolutely true. At the same time, these people have rent to pay and mouths to feed. They can't always afford to invest their time without some sort of tangible return. A promise of a percentage 'if and when' is great, but it doesn't pay the bills today. The old adage, 'you get what you pay for' often means that when people do 'favors' they don't give you their best work when other clients who can afford their rates are waiting in line.

Here are some very loose guidelines to keep in mind as you begin to build your budget to support your dream:

- A patent search will run from $250 to $1000, depending on the people you hire and the complexity of the search.
  - ☐ One of the companies I use frequently is Patent Search International **www.patentsearchinternational.com**. They weigh in at $250; patent attorneys may charge more.
- Having a patent attorney prepare the patent can cost anywhere from $2000 and up, depending on the complexity of the invention. This does not include the patent office filing fee (under $1,000) or the lawyer's fee to review and respond to the patent office 'office actions'. When the patent office gives you the green light, meaning that your patent has been admitted, you will be charged an additional 'notice of allowance' fee as well.

  [Author's Note: Why would you pay so much to a lawyer if a company like PSI charges less? Companies like PSI help you get the ball rolling at a lower cost, but when it comes to the business of filing your patent(s) your big idea deserves all the resources and protections available under the law. This is when you need a patent lawyer because this is what s/he is trained to do.]

- Your mock up or prototype can cost as little as $500 and as much as $5000 for a basic product.
  - ☐ More complicated concepts with electronic components and computer chips can cost considerably more.
- Engineering designs can run from $500 to $5,000, based on the fees the engineer charges and the time it takes to deliver the final designs.

Once you have your engineering design completed, there are two choices available to you as you face the next step:

1. You can take the greatest risks and reap the greatest rewards when you choose to do it all yourself, becoming the manufacturer, distributor and customer service guru all rolled into one.

- Tooling costs will depend on the number and size of the component parts and materials used or needed.

- Manufacturing and assembling your first run can cost as little as $2,000 and as much as $50,000, even more.

- Packaging, marketing, advertising, warehousing, distribution and staffing all come with their attendant costs as well.

2. You can choose the less expensive option of licensing, where someone else carries the heaviest burden and your investment is limited to the patent search, the patent filing costs, making the mock up and hiring the design engineer.

- Once the product is manufactured, packaged, distributed and sold, you earn a percentage of the revenue the licensee receives.

   DEFINITION: A licensee is a manufacturer with distribution channels who pays you a percentage. The percentage you receive is called a royalty.

Whether you choose to manufacture or license your product, you need to know how to develop a budget and set up a financing schedule to pave the way for the expenses to come. Building anything comes with fresh sets of expenses at various stages of development. Once you have a better idea of the costs that are involved, you can plan your financing wisely. The last thing we want to have happen is that you wake up one day and realize that your venture is under capitalized and your budget fails to support your needs.

As you move forward through this book and launch your dream, we want you to remember a very important thing: While the inventing process may be a love for you, the minute you involve anyone else in the process you are stepping outside the realm of personal dreams. You're asking people to invest their time and commit their talent in someone else's cause. Taking the time to familiarize yourself with the bottom line and place tangible dollar values on your experts' contributions will train you to ask yourself "What's in it for *them*" before you go looking for mentors and partners. We'll go into greater detail on the various costs involved in setting up and running a small entrepreneurial venture in Part Three.

Tools to get you up to speed.

Why would you re-invent the wheel when it's already out there waiting for you? To be informed and to fast-track your success, you need to leverage your knowhow by taking advantage of the educational material that is available to you.

In the Resource Guide located at the back of this book, you will find links to industry contacts, periodicals, magazines and books that will be valuable additions to your reference library.

PART TWO: GETTING STARTED

Part Two is designed to teach you to walk before you run.

No matter how brilliant your big idea is… no matter how desperately everyone needs it to enhance the quality of their lives, unless you take care of the details as and when they need to be addressed you are doomed to fail.

The chapters that follow are your blueprint for success. We can't urge you strongly enough to use it. Always remember… Fail to plan and you plan to fail.

CHAPTER FIVE: YOU HAVE AN IDEA!

NOW WHAT?

This may sound crazy, but the first thing to do when an idea enters your mind is get it out of there... translate it into tangible form as the first step in making it real.

The world is full of dreamers who never take this step. What sets you apart from them is the habit you create that takes your ideas from the dreaming stage and gives them substance, allowing you to retrieve and research, shape and share them to give them life. Today, there are many ways this can be done...

- You can keep a notebook handy;
    - ☐ Use a notebook that you can carry in your pocket or purse.
    - ☐ Use it! It won't do you any good if it stays in your pocket or purse.
    - ☐ Get in the habit of writing notes in simple point form that you can read and understand later on.
    - ☐ Write a keyword that helps jog your memory, followed by the trigger and the rough idea that has just popped into your head. As an example, if you've been wrestling with a packaging issue, write the word packaging and then a few words to remind you of the trigger and then describe your idea: [packaging graphic – Post It found in fridge - hand written Post It graphic on package]

- You can sketch or doodle it and keep it in a folder;
    - ☐ Keep your rough sketches and doodles where they're easy to find
    - ☐ Make labels for each folder, identifying separate stages or tasks and sort your drawings and comments into those folders
- You can send a text message to yourself;
    - ☐ Sometimes it's quicker to text a message than to pull out your notebook. Sending yourself a text message is an excellent way to focus your attention on a specific idea at a later time.
    - ☐ Use the same keyword reminders so that your brilliant idea doesn't end up being a big question mark later on.
- You can write an email;
    - ☐ If you're in the middle of some other work, possibly answering your morning emails, you can send a quick note to yourself without taking the time to get out your notebook [the one you keep leaving in your pocket or purse, remember?]
- You can record it on an audio file;
    - ☐ Audio files are excellent ways to make notes and save them. They have the added advantage of allowing you to say exactly what you're thinking without having to create short cuts.
- You can write it on a post-it note and stick it on the wall;
    - ☐ Sometimes the old technology is the best technology! Keep a pad of post it notes handy and create eye catching reminders that leave no tell tale marks when you take them away.
    - ☐ Once you've had a chance to sit down and think

about the note you've left, you can put the post it note in the folder you've made.

The method that works best is the one you actually use. As soon as you get an idea in your head, get it out of your head right then and there.

No matter how clever it is… no matter how impossible it seems that you could ever forget it… you can and you will. Unless you get it out of your head and put it someplace safe, it will get lost and forgotten…

- The phone rings; you answer it and get caught up in the conversation. When you hang up the phone, you stare off into space. You can't believe that your brilliant idea has vanished just like that.

- The car ahead of you runs a stop sign and hits someone. You'll pull over and see what you can do to help. When you get back in your car, will your brilliant idea be there waiting for you? Nope.

- How about when the water boils over on the stove… when the cat scratches the baby… when the car falls off its jack stands… when the computer crashes before you've had a chance to save your work… when Life continues to unfold… what will happen to your brilliant idea then?

Exactly. So choose the method that works for you and use it every time a brilliant idea enters your head. Sketch, doodle, write, text or record it out of your head as fast as it came in. Put it in a safe place so it will be there when you've jumped Life's latest hurdle and you're ready to start working on your dream.

Now what?

When Life cuts you some slack, you need to take a closer look at your brilliant idea and make sure that it's really yours. The simplest, most effective way to do an informal background check is to take advantage of the tools that are available to you free of charge on the Internet.

If you don't own a computer, you can visit your public library or an Internet café. You can use their computers to access the search capabilities of companies that offer this service at no charge to you. If you don't have a favorite or favorites, this is a shortlist...

- www.google.com
- www.yahoo.com
- www.msn.com
- www.ask.com

To find others, simply type in the keyword phrase "search engines" when you go on the Internet and choose from among the ones that appear after you click.

When you've picked a search engine, go to their website and follow their instructions on the best way to use their service. When you're ready to begin your search, try as many different word combinations as possible to describe your idea. [If you invented a shoe, you would try specific descriptors – men's, ladies, children's, sneaker, trainer, leather, formal, classic, sandal, orthopedic and so on – as well as categories like shoe and footwear] This is good practice and it will hone your patience skills.

The reason this step is essential and should be taken sooner rather than later is that you need to know if you've come up with something entirely new or your idea already exists exactly or in a different form.

Each of these results presents different challenges and you need to know what they are before you spend time, effort and money to develop your vision. Let's take a moment now to consider these scenarios...

☐ *Your idea is unique and entirely new*

- Congratulations! You've passed the first hurdle if your preliminary search fails to find a product or service like yours. You need to keep digging to make sure. We'll cover this in detail on the pages that follow
- You also need to find out *why* your product or service doesn't exist. While you may be the first with a brilliant idea, there

have been instances when further research has shown that a product or service was too impractical to use. We will cover this topic in more detail as well.

☐ *Your idea is similar but substantially different from others on the market*

- Well done! You may have found a way to build a better mouse trap! You need to keep digging to find out how many different models there are and how successfully they serve their market.

- You also need to decide if you want to devote your time and energy to develop, produce and sell an invention that already exists in a different form. You may not want to showcase your concept in this way.

☐ *Your idea is just like many others already on the market*

- Don't give up hope! If someone else is making money with an idea just like yours, it means that you are on the right track! You've got what it takes to create a winner and it's just a matter of time before you come up with something new.

- The market has plenty of room for competing products and services. You may want to pursue this one and dedicate time and energy to come up with a clever campaign to position yours ahead of the rest.

Now what?

As an inventor, you will need to familiarize yourself with the United States Patent & Trademark Office. You can find them online at **www.uspto.gov**. This is the government agency you will contact to register your concept. You will be able to take advantage of many valuable services that are available on their website.

One of these services is 'pre-screening'. Once you reach their website, simply click on "patents" and use their search service just as you used the commercial ones provided on the Internet. The Patent & Trademark Office will have the answers you are looking for.

Now what?

You want to name your invention. Giving it a name makes it easier to talk about and makes it feel real to you. When you take the time and effort to name it, you are committing to it in an entirely new way. In many ways, your invention is your baby. It came from you and it deserves a name that fills you with pride. When you encounter red tape hurdles and technical setbacks, the excitement that is fueled by pride of ownership keeps you focused on your goal.

Let's say you invented a reversible all-weather dancing shoe. [Yes, we think it's pretty silly too… it's just an example.] Do you want to use that phrase every time you talk about it? Suppose you name it using the first letter from each word: RADS. The name is much easier to use and when the shoe has a name, it seems more like a real product. You can visualize the packaging and promotional concepts more clearly. It feels more real and you feel much more committed to its success.

Even if you love it, the first name you think of may not be the one you end up using. You need to search to make sure the name you've come up with isn't already being used. You will want to search domain names as well [A domain name is the name given to a website.] and you can use **www.whois.net** to do that. The 'who is' site gives you the names and contact information of the people who have registered their domain names. This makes it easier to contact them and find out if they're really committed to the name or might be willing to let it go for a nominal fee.

Position the name you are considering as a website name [as an example, www.rads.com] and let the Internet tell you whether someone is already using it. It may be registered as a 'dot com' business by someone who sells radiators. In that case, you might try to make the name more specific [www.radshoes.com].

If the name isn't being used, you can register your domain name very easily and inexpensively through a domain registry business. You

You And Your Big Ideas

can search and compare registry companies like www.**godaddy.com** and **www.register.com** to register your domain inexpensively. Shop and compare to get the prices and services that suit your budget and your needs. Search for 'domain registry' businesses to get several prices and compare.

Now it's time to take a quick trip back to the Patent & Trademark Office website [**www.uspto.gov**]. This time you want to go to the Trademark Section and check your proposed name and product categorie(s) to find out if anyone has registered the name you are planning to use as your trademark. If no one has registered your name, it's time to move to the next step on your checklist:

Before you click, register and buy a name, go through the process at least two more times. When you've got three names that you really like, it's time to do some informal market research: Run the names past ten people whose opinions you value and ask them to give you their feedback by listing them in the order they like best and telling you why they made those choices..

If you feel uncomfortable talking openly about your idea, you can draw up a simple Non-Disclosure Agreement. Agreements like this are available online. Try **www.asktheinventors.com** and **www.score.org** to get a Standard agreement and modify it to suit your needs. Keep it simple and share only as much information as necessary in order to provide a basic understanding of your concept and the market you think it will serve. Services like SCORE – run by retired executives – provide many similar services to entrepreneurs.

When you approach people to give you their feedback, you need to make it clear that you're looking for their honest reaction. People who say kind things to save your feelings aren't helping you in the long run. So the onus is on you to choose individuals who understand that you're coming to them for a fresh perspective and that their input has value to you. Sometimes a casual suggestion to tweak it this way or position it that way can make all the difference when you take your invention to the next level.

CHAPTER SIX: PROTECTING YOURSELF

A patent search provides you with absolutely essential information. Even though a quick search may show that you're in the clear for patent protection or that you have an original idea, you MUST protect yourself. Sooner rather than later, you need to have a patent search done.

As I've already mentioned, one of the companies I use is Patent Search International. My contact there is Ron Brown. He charges $250 for a patent search and for that fee you get a patent search, an opinion letter from a patent attorney and a guarantee. The opinion letter is an extremely valuable tool that references many of the ideas and prior art currently within the USPTO database that may be similar to yours. You can use the search findings information included in this search when you file your patent application.

What is a claim?

Claims listed on a patent are very important. You want to try to have as many claims listed on your patent as possible so that no one can have a product that overlaps yours. Claims define the scope or boundaries of the invention in much the same way that a deed defines the land boundaries of real property. They are the documentation that a court examines to determine if infringement of a patent has occurred.

I've dealt with many excellent patent attorneys with regard to claims. According to Richard Klar, who was kind enough to provide the following definition, "The language of a claim for a utility patent [e.g.: a patent protecting functional features] is preferably written in general terms with minimal use of specific descriptors so that it covers the broad concept of the invention and a potential infringer does not avoid infringement by eliminating one element or one step in a process. Thus, if you claimed a product having elements A, B and C, a competitor making a predicate with elements A and B only automatically avoids infringement. To protect the specific elements or steps, your minimally worded broad claim would have extra claims added that refer specifically to each additional feature."

[Author's Note: I went to the USPTO site to get their official wording for this next part because it's important to understand how the law interprets these terms. This is a heads up that the language makes it a bit of an uphill read, but since you're going to need to get familiar with a number of official procedures as an inventor, this is a good place to start. By the way, you can find a list of patent attorneys and agents under the RESOURCES link at the top of the home page.]

What is 'patentable'?

In order for an invention to be patentable it must be 'new' as defined in the patent law, which provides that an invention cannot be patented if:

"(a) the invention was known or used by others in this country, or patented or described in a printed publication in this or a foreign country, before the invention thereof by the applicant for patent," or

"(b) the invention was patented or described in a printed publication in this or a foreign country or in public use or on sale in this country more than one year prior to the application for patent in the United States . . ."

If the invention has been described in a printed publication anywhere in the world, or if it was known or used by others in this country before the date that the applicant made his/her invention, a patent cannot be obtained.

If the invention has been described in a printed publication anywhere, or has been in public use or on sale in this country more than one year before the date on which an application for patent is filed in this country, a patent cannot be obtained. In this connection it is immaterial when the invention was made, or whether the printed publication or public use was by the inventor himself/herself or by someone else.

If the inventor describes the invention in a printed publication or uses the invention publicly, or places it on sale, he/she must apply for a patent before one year has gone by, otherwise any right to a patent will be lost. The inventor must file on the date of public use or disclosure, however, in order to preserve patent rights in many foreign countries.

Even if the subject matter sought to be patented is not exactly shown by the prior art, and involves one or more differences over the most nearly similar thing already known, a patent may still be refused if the differences would be obvious.

The subject matter sought to be patented must be sufficiently different from what has been used or described before that it may be said to be non-obvious to a person having ordinary skill in the area of technology related to the invention. For example, the substitution of one color for another, or changes in size, are ordinarily not patentable.

Who can file a patent?

Attorneys at law and persons who are not attorneys at law can file a patent. The former persons are now referred to as "patent attorneys" and the latter persons are referred to as "patent agents." Both patent attorneys and patent agents are permitted to prepare an application for a patent and conduct the prosecution in the USPTO.

Patent agents cannot conduct patent litigation in the courts or perform various services which the local jurisdiction considers as practicing law. For example, a patent agent could not draw up a contract relating to a patent, such as an assignment or a license, if the state in which he/she resides considers drafting contracts as practicing law.

What Is a Patent?

A patent for an invention is the grant of a property right to the inventor, issued by the United States Patent and Trademark Office.

Generally, the term of a new patent is 20 years from the date on which the application for the patent was filed in the United States or, in special cases, from the date an earlier related application was filed, subject to the payment of maintenance fees.

U.S. patent grants are effective only within the United States, U.S. territories and U.S. possessions. Under certain circumstances, patent term extensions or adjustments may be available.

The right conferred by the patent grant is, in the language of the statute and of the grant itself, "the right to exclude others from making, using, offering for sale, or selling" the invention in the United States or "importing" the invention into the United States.

What is granted is not the right to make, use, offer for sale, sell or import, but the right to exclude others from making, using, offering for sale, selling or importing the invention.

Once a patent is issued, the patentee must enforce the patent without aid of the USPTO.

There are three types of patents:

1) <u>Utility patents</u> may be granted to anyone who invents or discovers any new and useful process, machine, article of manufacture, or composition of matter, or any new and useful improvement thereof;

2) <u>Design patents</u> may be granted to anyone who invents a new, original, and ornamental design for an article of manufacture; and

3) <u>Plant patents</u> may be granted to anyone who invents or discovers and asexually reproduces any distinct and new variety of plant.

Application For Patent: Non-Provisional Application for a Patent

A non-provisional application for a patent is made to the Director of the United States Patent and Trademark Office and includes:

(1) A written document which comprises a specification (description and claims), and an oath or declaration;

(2) A drawing in those cases in which a drawing is necessary; and

(3) Filing, search, and examination fees. Applicant must determine

that small entity status is appropriate before making an assertion of entitlement to small entity status and paying a small entity fee. Fees change each October. The fee schedule is posted on the USPTO Web site.

All application papers must be in the English language or a translation into the English language will be required along with the required fee set forth in 37 CFR 1.17(i). All application papers must be legibly written on only one side either by a typewriter or mechanical printer in permanent dark ink or its equivalent in portrait orientation on flexible, strong, smooth, non-shiny, durable and white paper.

The papers must be presented in a form having sufficient clarity and contrast between the paper and the writing to permit electronic reproduction. Each document in the application papers must all be the same size - either 21.0 cm by 29.7 cm (DIN size A4) or 21.6 cm by 27.9 cm (8 1/2 by 11 inches), with a top margin of at least 2.0 cm (3/4 inch), a left side margin of at least 2.5 cm (1 inch), a right side margin of at least 2.0 cm (3/4 inch) and a bottom margin of at least 2.0 cm (3/4 inch) with no holes made in the submitted papers. It is also required that the spacing on all papers be 1 1/2 or double-spaced and the application papers must be numbered consecutively (centrally located above or below the text) starting with page one. The specification must have text written in a non script font (e.g., Arial, Times Roman, or Courier, preferably a font size of 12) lettering style having capital letters which should be at least 0.3175 cm (0.125 inch) high, but may be no smaller than 0.21 cm (0.08 inch) high (e.g., a font size of 6). The specification must have only a single column of text.

The specification must conclude with a claim or claims particularly pointing out and distinctly claiming the subject matter which the applicant regards as the invention. The portion of the application in which the applicant sets forth the claim or claims is an important part of the application, as it is the claims that define the scope of the protection afforded by the patent. The claims must commence on a separate physical sheet of paper.

More than one claim may be presented provided they differ

from each other. Claims may be presented in independent form (e.g. the claim stands by itself) or in dependent form, referring back to and further limiting another claim or claims in the same application. Any dependent claim which refers back to more than one other claim is considered a "multiple dependent claim."

The application for patent is not forwarded for examination until all required parts, complying with the rules related thereto, are received. If any application is filed without all the required parts for obtaining a filing date (incomplete or defective), the applicant will be notified of the deficiencies and given a time period to complete the application filing (a surcharge may be required)—at which time a filing date as of the date of such a completed submission will be obtained by the applicant. If the omission is not corrected within a specified time period, the application will be returned or otherwise disposed of; the filing fee if submitted will be refunded less a handling fee as set forth in the fee schedule.

The filing fee and declaration or oath need not be submitted with the parts requiring a filing date. It is, however, desirable that all parts of the complete application be deposited in the Office together; otherwise each part must be signed and a letter must accompany each part, accurately and clearly connecting it with the other parts of the application. If an application which has been accorded a filing date does not include the filing fee or the oath/declaration, applicant will be notified and given a time period to pay the filing fee, file an oath/declaration and pay a surcharge.

All applications received in the USPTO are numbered in sequential order and the applicant will be informed of the application number and filing date by a filing receipt.

The filing date of an application for patent is the date on which a specification (including at least one claim) and any drawings necessary to understand the subject matter sought to be patented are received in the USPTO; or the date on which the last part completing the application is received in the case of a previously incomplete or defective application.

Provisional Application for a Patent

Since June 8, 1995, the USPTO has offered inventors the option of filing a provisional application for patent which was designed to provide a lower cost first patent filing in the United States and to give U.S. applicants parity with foreign applicants. Claims and oath or declaration are NOT required for a provisional application. Provisional application provides the means to establish an early effective filing date in a patent application and permits the term "Patent Pending" to be applied in connection with the invention. Provisional applications may not be filed for design inventions.

The filing date of a provisional application is the date on which a written description of the invention, and drawings if necessary, are received in the USPTO. To be complete, a provisional application must also include the filing fee, and a cover sheet specifying that the application is a provisional application for patent. The applicant would have up to 12 months to file a non-provisional application for patent as described above.

The claimed subject matter in the later-filed non-provisional application is entitled to the benefit of the filing date of the provisional application if it has support in the provisional application. If a provisional application is not filed in English, and a non-provisional application is filed claiming benefit to the provisional application, a translation of the provisional application will be required. See title 37, Code of Federal Regulations, Section 1.78(a)(5).

Provisional applications are NOT examined on their merits. A provisional application will become abandoned by the operation of law 12 months from its filing date. The 12-month pendency for a provisional application is not counted toward the 20-year term of a patent granted on a subsequently filed non-provisional application which claims benefit of the filing date of the provisional application.

A surcharge is required for filing the basic filing fee or the cover sheet on a date later than the filing of the provisional application.

Legal Fees

Patent agents and attorneys charge between $100/hour to $500/hour to draft a licensing agreement.

[Author's Note: Okay, we're back. There's no getting around the fact that legal wording is slow going. I like to think of it as preventative medicine: it may not be fun to take in, but it performs an important function that protects you from worse things than a mild headache from reading convoluted phrases.

Let's get back to the simple wording, shall we?]

CHAPTER SEVEN: MAKING IT REAL

Congratulations! You've made it through all the headache-generating language and you've successfully completed the patent search process. Well done.

One of two things (other than the headache) has happened: You've learned that you either have a stand-alone, patentable idea or you don't.

One

When I get my patent search back and it is confirmed that I have a patentable item, I get a <u>provisional patent</u> – this can cost anywhere from $600 to $1000, including the filing fee of about $100, to have an attorney put together a Provisional Patent application for you.

A Provisional Patent gives you a year of protection for your invention (see the legal section in the previous chapter for the formal description of a Provisional Patent). During this time you can develop, market and hopefully sell your idea and be patent protected. At the end of the year you will have to go back to the Patent Office and file an application for a Utility Patent or a Design Patent – this is the one that will protect you for the next 20 years.

[Author's Note: A design patent protects a design you may have developed, but it doesn't offer as much protection. A very slight change in the design allows someone to patent their own idea and you will not be protected. Unless you're a botanist, a plant patent won't be of much use to you.]

The provisional patent is an excellent route to take for start up inventors because it's inexpensive and relatively quick. The reason I advise you to use a patent attorney rather than a patent agent is because the Utility Patent (a.k.a. the Non Provisional Patent) that you might file for later will be based on the wording and filing date in the Provisional Patent. The last thing you want to do is skimp on the protective measures you undertake to keep your invention safe.

Two

If the search comes back indicating that my idea is not patent-able, I can still go ahead and file a provisional patent. Why would I do this? It gives me the Patent Pending status that enables me to go out and talk about it.

Why do I want to go out and talk about it if I can't patent it? You may be able to trademark the name and license it along with a modified version of the product. Licensing the name can provide residual income in the form of royalties just the same way licensing the actual invention does. A trademarked name can become your most valuable marketing tool because it can allow you to brand your idea.

In marketing circles, 'branding' is what happens when the name of a product or service is recognized more readily than the individual product or service itself.

Nike is a classic example of branding. People have come to equate exceptional quality and peak performance with the Nike label thanks to the endorsements made by celebrity athletes around the world. As a result, athletic clothing and accessories that carry the Nike label also carry prices that far exceed the actual value of the products themselves. The Nike trademark generates residual income in the form of royalties paid to the licensor who patented the trademark.

We've already talked about getting a name and searching to make sure that the related website domain name is available. How do you go about getting a Trademark? You go back to the U.S. Patent & Trademark office online and follow the steps they outline. Alternatively you can visit your public library.

Once you've completed the patent and/or trademark steps and you're ready to move forward, the next thing you need to do is take your idea off the page and make it work.

CHAPTER EIGHT: MAKING IT WORK

Congratulations! Your idea was given a green light when you did your patent search. You've taken out a provisional patent to protect yourself and now it's time to move on to the next step in the process.

The first thing you have to do is make a proto-type – a working model of your invention that allows you to share your idea with experts and get their input before you move forward.

A simple and inexpensive way to do this is to visit stores like Walmart and Target and any discount craft stores that you know. You're looking for materials that you can adapt to give your invention tangible form. As we discussed briefly in Chapter 4, the costs can be minimal for a basic model but often go higher in direct proportion to the complexity of the product.

Whatever you do, DON'T RUSH THIS STEP! Make a prototype that will capture the imagination of everyone who sees it. Make sure it is functional as well. Double check your cost and budget estimates.

Once you have the working model done, you need to sit down with someone who can help you figure out the mechanical, technical and cost-of-production details. This person is known in the industry as a engineer. There are many different types of engineers – mechanical, design, electrical, etc – and you will need to do your homework to find the one that's right for you.

[Author's Note: This is also a budgeting issue: when you allocate funds for each step in the process, the design fees fall into the cost-of-production category.]

My best advice is to go in search of engineers who work on products like yours and ask them if they're willing to take a look at your concept. Be sure to ask if they are interested in taking on the job themselves and ask for a quote on both the design and development aspects. Never go with just one quote and always ask for their input:

- Any suggestions they may have to improve on your idea
- Any contacts they may have who can take the proto-type to the next level.
  - ☐ They may know proto-type houses
  - ☐ They may have contacts with manufacturing entities here and overseas, especially if they've been involved in previous completed inventions.

In this way, you are considering the design engineer as a potential partner and mentor. You can invite them to become a partner, be involved in your business plan and offer a percentage as a benefit to them – this can include royalties from licensing and/or percentage of sales of the product. If the design engineer gets involved, you will have the added advantage of their knowledge and expertise when it comes to quality controls and specifications. Offer to provide a Non Disclosure Agreement that all parties sign.

Fees

Engineers typically charge from $50 to $400/ hour or more. Although it is impossible to quote the exact number of hours it may take them, you can expect an invoice that covers a minimum of 10 hours' design work.

What to expect from an engineer.

- The engineer works up various alternative solutions on paper to discuss with you as a first step to find out if it will work and how it will work.
- This information will be applied as the engineer translates your concept into engineered drawings that you can take to a factory/manufacturer to make a prototype and/or go directly into production.

- The engineer usually leaves room for modifications that may need to be made, either to allow for your design changes or to comply with concerns being raised by the factory or the manufacturer.
- The engineer should make himself available to you if you have questions that are being asked from the factory that the engineer will know how to answer.

Which comes first?

When I've had my AHA! Moment, the first thing I always do is make sure I've come up with something worth pursuing. I run the Google and Domain search engines and then I get a patent search done. Then I create the mock up or proto-type. If the idea is patent-able, I consult with an engineer to make sure that my idea can be manufactured cost effectively before I file the Provisional Patent.

I do my due diligence when I pick a design engineer… I take a look at their previous projects… get second opinions… talk to my mentor… trust my gut instinct and make sure that my invention is protected before I give too many secrets away.

When I have the drawings and the provisional patent, my next step is to see if I can interest someone in licensing the product or the name. If the idea is patent-able, having the engineered drawings already made enhances my stature and the credibility of the product to the potential licensee.

Before making presentations to potential licensees, I always ask them about their royalty rates and if they pay advances (monies up front to recoup some or all of your money put into the idea up to this point) and/or guarantee of money to be paid for the rights to use your idea is included: How much of a percentage they are prepared to pay me when the item sells. You need to make a checklist for yourself when you do this.

1. How can you tell if their rates are fair?
    - Ask them for names or products of other licensors.
    - Ask other licensees and compare.

[DEFINITION REVIEW: A licensor is the person who owns or invented the product; the licensee is the person who pays for the right to sell the product.]

2. How much does your industry pay in royalties?
   - I have come across royalty fees offered from 3% to 7%.

3. Is your industry a high royalty or low royalty industry?
   - Ask other inventors what they earn
   - Research royalty rates from industry licensing handbooks like *The Licensing Business Handbook* (EMP Communications)

4. What are the terms and conditions that apply?
   - Licensing contracts can last for different periods of time. The specific terms and conditions that you agree to are part of the negotiation process.

5. Who can help you make the decisions that are right for you?
   - If you have a licensee interested in licensing your product or your brand name, hire a patent attorney to draw up or review the proposed agreement and help you in the negotiations.

If you have the time to make the phone calls, attend the trade shows, make your own contacts, negotiate the deals and prepare the agreements once you have a taker for your product, you can do it yourself. But for many people, the idea of picking up the phone and making presentations fills them with dread.

If that is how you feel, there's good news. Your discomfort does not have to stop you from realizing your dream. You can easily hire a licensing agent and have him or her do the hateful work on your behalf. If you want someone else out there instead of you, working their contacts to find the best partner/licensee for your product, a licensing agent can do it all.

You still have to make some phone calls to educate yourself about licensing agents. Find out…

- What fees they charge

- How much of a share do they take when they find a partner/licensee
- How long they will be representing you
- How much follow up you can expect and when you can expect reports on their progress
  - I have not met a licensing agent yet who represented just one project for one client. Generally they have several projects on the go at the same time, focusing their attention on what's hot and pushing aside what's not until they have time to give it the attention it needs to make it hot.
- What expenses you may have to incur
  - Attending trade shows,
  - Making presentations,
  - Airfare, accommodations, per diem, etc.

So, why would you use a licensing agent?

- Licensing agents do not add to your financial burden up front.
  - They earn their money in the form of percentages they take from your royalty rate. It is in their best interest to negotiate higher royalties because they get more money right along with you.
- They save you time and time is money.
  - They know things it could take you years to figure out.
  - They make contacts to promote your product.
  - They ferry contracts back and forth for your signature.
  - They make sure that your royalty checks are paid.

Okay, let's say you hired a licensing agent who did everything right and you have licensed your idea. The licensee is manufacturing and marketing and distributing your product. You are collecting royalty checks every quarter and there is no risk to you. Is that the end of the story? Of course not!

While all this is going on, you can look to the future. You can use the knowledge you have acquired and get quotes to find out what it would cost to cut out all the middle men, take on the risk and produce your product yourself. You can take your original engineered drawings to other manufacturers to get quotes. You can talk to distributors and retailers. You can talk to your mentors and other inventors who have walked a mile in your shoes. When the term of the licensing agreement is about to expire, you have a choice: You can renew your current agreement, with or without your licensing agent's involvement, or strike out on your own and go into business for yourself.

PART THREE: THE FORK IN THE ROAD

The inventing process isn't a simple one. Time and cost often hold creators back.

At this point you may have to start making some practical decisions, especially if your finances are limited. You may need to start looking for partners who are willing to invest their time and money in your innovative concept and in you.

Whether you decide to go it alone or take on partners, you need to create and follow a business-like formula that will allow you to keep a clear head and maintain control as the days, weeks and months unfold.

Anyone who has ever embarked on a new venture will tell you that the greatest challenge they faced was the learning curve. Without exception, people under estimate how long it will take them to get things done. Without exception, their ignorance will drive them to distraction. Without exception, they will find that they have put the cart before the horse. Without exception, they will tear out their hair as they retrace their steps to do it right. And so will you.

CHAPTER NINE:
TAKING CARE OF BUSINESS

Think of your invention as a newborn child. Even if you've never been blessed with children, you know from the stories your friends and family have shared that it takes a lot of teamwork and planning to pilot them safely through their fragile formative years and get them ready to go out into the world.

Whether you choose to manufacture or license your invention; whether you choose to go it alone or build a team to support you, the first thing you have to do is build a solid framework that will allow you to enjoy the process that is unfolding before you.

GETTNG STARTED

If you have absolutely no money above and beyond your basic living expenses, it is unlikely that you will be able to turn your invention into a commercial product unless you are willing and able to raise or borrow the necessary capital to get started.

Whether you are licensing you idea or you plan on bringing it to market yourself, you are going to need some money to get you going. At the very least, you will need to have the funds in the bank to pay for your patent search, make some drawings, make a prototype and file for some protection. You're also going to need enough money to get your business started. Even if you plan on operating on a shoestring budget at the start, you still need to pay for things like your incorporation, your bank account, your business cards and letterhead, a computer and an internet connection if you don't already have one and a long distance phone plan.

If you're looking at licensing because you're thinking that, as a licensor, you won't need to invest in your idea, I can assure you that this is a wrong assumption. Prospective licensees will want to see how much of *your* money you're wiling to put into your idea because this tells them how committed you are to its success. If they are going to bring your product into production and distribution and take on all that risk, they want to know that you are equally invested in your idea's long term success.

If youre planning on keeping the controls in your hands and manufacturing your idea yourself, you will need to pay the contract manufacturer – the provider you hire to deliver the product. You will need to invest your time to get a number of quotes that will tell you how much money will be needed for tooling and a first production run. Even if the manufacturer is willing to give you discount incentives, it's still going to be a sizeable investment on your end. You're also going to need money to get your sales and distribution teams in place. You'll need to rent a facility and furnish it and hire staff and set up payroll processes. All of these steps will cost money long before your idea is ready to generate any kind of revenue stream.

How Much Is Enough?

Someone with experience will be able to tell you how much money you will need to fund your project. A good source is your mentor, of course, because s/he has been through the process many times before. You can also get in touch with fellow members of the inventors' group(s) you belong to. Ask them how they financed their inventions and what it cost them to complete the many processes and steps and use this information to get ideas on how you can do it.

How will you fund this venture?

No matter where you go for funding, the key to success is to conduct yourself in a businesslike manner. You need to be professional at all times – in your phone call inquires, in your letters and emails and in your presentations.

You And Your Big Ideas

1. Friends and Family

Many novice inventors make the mistake of thinking that their personal networks are their best sources of funding support. In fact, imposing such a heavy burden on personal relationships can be the very worst thing you do. Your friends and family are your *personal* supporters. Their loyalty and commitment are invested in *you*. That their love is unconditional does not mean that it is a transferrable asset. *'If you loved me, you'd invest in my dream'* is not a reasonable assumption to make because it imposes unfair demands on the relationship. Your dream *isn't* their dream unless and until *they* decide to make it theirs.

If you do decide to approach people in your inner circle, treat them with the same respect that you would give to a stranger. In other words, do it on a businesslike basis. Make a business proposal that clearly defines the risks and rewards that come with the investment you are asking them to make. Work out percentages they will own that entitle them to share in any future success. Work out realistic repayment plans that clearly explain how and when they will get repaid if your idea fails. Don't cut corners. Get professional legal advice and draw up a Non Disclosure Agreement and a simple contract between both parties. Show them in tangible ways that you respect them. Take pro-active steps to protect their investment even before they make it, to prove that you have their best interests at heart.

2. Yourself

As you put time and energy into your Big Idea, you really become attached to it. In your excitement and enthusiasm, it is very easy to make the decision to put your hard earned money into it as well.

Before you decide to cash in your savings or take a percentage of your income each pay period and put it toward a project fund, you should discuss this decision with the people who depend on you. As suggested in the previous point, it is very important to explain your idea and the opportunity it represents to your loved ones in businesslike terms. Giving them the respect you would give to a stranger will help your wife, girlfriend, husband, boyfriend, parents, etc. to see your plan as an investment in everyone's future rather than a selfish whim.

Whatever you do, don't hide what you're doing from them. Don't take money out of the budget that keeps the roof over their heads and food on their plates to take a chance on your dream.

3. Small Business Administration Bank Loans

Many inventors get themselves going by going to their local branch of the Federal Government's Small Business Administration and asking them for help in securing a loan from a local commercial bank.

Because these SBA-supported initiatives are usually collateralized by a home or some other assets you may have, this means that you will need to get your family involved. Even with SBA approval, the amount you can borrow is justified by how much collateral you are putting up: collateral that, in many cases, is an asset that is jointly owned by you and your spouse.

4. Government Grants

Grants are also available for specific inventions and are given from government departments for specific inventions. For example, the Department of Energy may be offering a grant to support ideas and innovations that protect our environment or conserve energy. Visit your local government office to find out more about what may be available and fit your invention.

5. Venture Capital

Venture Capital lenders have money to lend and it could be yours after you have provided the necessary proofs that your product would be or is viable in the marketplace. Traditionally, venture capital is part of the second or third stage of a start up, coming in after you have invested your own resources to get the project off the ground. Often, venture capital firms are interested in investing in ventures that need tens of millions of dollars of capital to reach their potential. **Pratt's Guide** is an excellent source of information on venture capital firms.

Venture Capitalists can have great connections to help build momentum for your company. Moves you plan on making with your product most likely will have to be run by your VC group for review and/or approval. This is an excellent opportunity to take advantage of their experience, contacts and know-how. They will typically take a good percentage of your company and percentage of your profits in exchange for their support.

6. Angel Investors

An Angel investor is someone who contributes based on their belief in your business plan, your big idea and most importantly, YOU. Angels are often said to invest emotional money, while venture capitalists invest logical money. Angels typically protect their investments by getting involved in the management of the venture. The best angels bring a lot to the table, investing their money, their time, their contacts, expertise, credibility and power. Angels typically keep a low profile and are hard to find unless you know how to access angel networks or clubs.

7. Other Sources

Be sure to look for announcements in places like Inventors Digest and in your local Inventor groups that call for new products. For example, many companies, commercial retailers and TV shows (like Everyday Edisons, as an example) often hold contests. If they select your idea, they can offer you a cash prize and/or royalty terms. Winning a contest like this can give you national exposure and make your idea more appealing to potential investors. Be sure to read the fine print when you sign up for these contests to be clear on what you may receive and what you may have to give them in return.

8. Last Resorts

Using credit cards, savings accounts, second mortgages and personal loans are last resort options. The rule is, "Do not borrow more money than you can make payments on indefinitely."

[Author's Note: For more information on funding sources, visit the following: http://inventors.about.com/od/fundinglicensingmarketing/a/inventionfunds.htm and www.webpatent.com, among others.]

SETTING UP YOUR BUSINESS

Your business needs an identity. You need the freedoms and protections that are available to a business owner. Taken together, this means that you need to establish a business and create the rules under which it operates.

Small Business support groups and agencies exist in many branches of government. Their sole function is to offer advice, training and support to individuals who are about to enter the exciting world of the entrepreneur. In many cases their materials are free and in some instances they can help you with start up essentials like business plans and loan applications. Some of them even offer grants.

- It costs money to create a legal entity and start a business, but there are ways to save money if you do your research and ask the experts.
  - ☐ Speak to an attorney and/or accountant and find out which is the best way to protect your ideas, yourself and your assets for income tax purposes.
- If you use your own name for the company, you're less likely to have competitors.
  - ☐ When choosing a name for your business, use the Google and Domain search engines just as you did when you were considering names for your product.
  - ☐ Make the list of your top three names and ask your friends to rank the names and tell you what factors influenced their decisions.
- You will want to get business cards and letterhead printed.
  - ☐ Keep the design simple and limit the colors to keep your costs down.
  - ☐ Remember that you will need to fax documents, so choose layouts and colors that translate well in black and white.
  - ☐ A business card can show that you're serious about your idea or you're not. Don't cut corners with the tool you will use to sell yourself and your ideas to future partners, investors

and industry professionals. You don't have to spend a lot of money to get a good design that accurately represents you. Inexpensive, good quality printing is available in small quantities at office supply stores and at any number of locally owned and operated printing houses.

- Even the smallest business can drown in paperwork, especially with all the research and filing multiple applications with various government agencies.
    - Invest in a business management course or attend classes offered by the small business agency to learn how to set up user-friendly procedures to record and retrieve your work.
    - Invest in the right tools:
        - A computer
            - You need a computer that isn't being used by the family for its various needs – recipes, family emails, homework and games.
            - You need a reliable email account and you need to store emails in specific folders so you can retrace your steps.
            - Always keep a back up copy of your emails.
            - Computers do crash. Save frequently using external methods.
            - Print your address books and keep them on file.
        - A scanner
        - A filing cabinet
        - Folders, binders labels
        - A daily journal or agenda.
    - Set aside a separate space that affords you the privacy you need to focus exclusively on your work. You cannot run a business if your children are using your office; or the washer/

dryer makes too much noise; or you have to clear away your paperwork to set the table for dinner.

- ☐ Set aside a time on a daily basis when you can work. Make sure that everyone knows this daily commitment is an essential service you must provide to your business if you want it to succeed.
- ☐ Create a data base of potential providers and support services. Update them regularly and make sure that the contact person has not changed.

• You will need a website. A website is a virtual office that everyone who uses the Internet can visit, 24/7. To set up your site you can hire a web designer or use a service like **www.godaddy.com**, which offers a range of inexpensive, comprehensive services from domain names to startup websites.

- ☐ Register a domain name to establish an identity on the Internet
- ☐ The name of your product should be registered and you may want to incorporate the name into your company's domain name.
- ☐ Dot Com names are the Level One names. You may also want to register the secondary alternatives [dot net, dot biz, dot org]
- ☐ Keep the name short so that people can remember it. Since the words will appear with no spaces and no capital letters, you want to pick a name that doesn't confuse the eye – words that begin and end with vowels or with repeated consonants are difficult to understand [Examples: successsells, miniinstructions, safeedits, etc]
- ☐ If the name is already taken, go to **www.whois.com** and find out who has registered it. If they're not using the name, you may be able to buy the domain name from them.
- ☐ Web designers usually have their name at the bottom of most web pages. If you like the work they've done for someone else's site, you've already taken the first step.
- ☐ Try to find a designer who will help you integrate new content quickly and easily. Your worst nightmare is a site

you cannot update because your designer has no appreciation of the urgency or has no commitment to you.

- ☐ If you can get a site whose software allows you to update and manage it yourself, you need a web designer who is willing and able to help you make the end result look professional.

- ☐ Don't make the mistake of loading up your site with unnecessary special effects. These do nothing to help communicate your message to the world and in many cases they slow the site down. The last thing you want is a site that keeps people waiting so long that they give up and leave.

- To set up a business phone, you can call your local telephone company to add a line. Another add on service is to get a toll free number and go one step further and capture an easy vanity number for your business or product go to **www.att.com** and use their toll free look-up tool and either reserve it with them or your local phone carrier.

  - ☐ Your vanity toll free number(s) can point to your local number. You can use this number in your advertising so people can contact you

    a) Free of charge, and

    b) Without geographic limitations implied or imposed

  - ☐ Get a virtual receptionist. A virtual receptionist is a traffic management tool that allows you to establish a professional image without the burden of paying someone to answer your phones and direct your calls.

  - ☐ If you have voice mail, make sure you check it frequently and clear the mailbox. Nothing is more annoying than hearing a recording that says 'mailbox is full'. Check messages frequently; return calls promptly.

- Dress for success. How you package and promote yourself can go a long way to build confidence in your innovative idea as well.

- Spend your money on your idea, but present yourself in a businesslike manner. Always be ready to receive a business client. Does this mean a suit and tie? No, of course not. It means clean, tidy, organized and professional. Ask yourself when you look in the mirror, would I invest my time and money in someone dressed like that?

CHAPTER TEN:
IF YOU DECIDE TO MANUFACTURE

This can be time-consuming and at the same time, enormously profitable...

- Taking on the role of manufacturing your invention carries the most risk and the greatest reward.
    - ☐ The most risk means that the buck stops with you. You will have to set up a business and hire specialists to take care of the following:
        - o Patent searches
        - o Patents
        - o Prototypes
        - o Engineering
        - o Manufacturing
        - o Marketing
        - o Packaging
        - o Warehousing
        - o Sales
        - o Distribution
        - o Shipping
        - o Invoicing – accounts payable and receivable
        - o Insurance
        - o Customer service
    - ☐ These variables can weigh in as costs that are high at the beginning; but once you have them in place it can be very, very rewarding for you.

- You must be an innovator and an entrepreneur.
  - In addition to coming up with your big idea, you also have to know how to set up and run the business to support it
  - You have to know your product and the industry it serves
  - You have to wear many hats, sometimes working a full time job to collect your paycheck while you do your due diligence to make sure that the life of an entrepreneur is right for you.
- When taking your idea from the ground and bringing it to this level, your passion for your product will work for you.
- You get to know your product inside and out
- You work closely with your factory with all the specs and samples, from packaging to finished product.
- Tooling will be a big expense [$5,000 - $100,000+ depending on the product]
  - In the US and Canada, the cost of tooling can be 30 - 80 per cent higher than in a place like China.
  - The complexity of the product will determine the cost of manufacturing the metal mold that gets filled with the material that will become your product.
  - As an example of surprises when it comes to tooling costs, I was getting quotes from various tooling manufacturers in the US anywhere from $30K to $90K. The quote I got from China was $9K. Don Debelak talks about "turbo outsourcing": finding a manufacturer to help with the tooling costs by absorbing the cost into the piece price. I was able to reduce the $9K quote by a third by increasing the cost per piece to start until the tooling was paid off. If you have guaranteed orders the manufacturer sees, they can make the production run happen with little or no money out of pocket.
- Before deciding to manufacture, find the right factory and partners.
- You must be willing to communicate with the factory from start

of production till finished product, meeting deadlines, shipping, warehousing, accounts payable/receivable, marketing, servicing, etc.

- Manufacturing is all about taking all the risk in order to reap all the rewards.

☐ ***Learn about your industry, find the right partners and make it happen!***
- Do you want this to be your full time job?
- Are you prepared to deal with manufacturers' jargon, inventory controls, shipping and receiving, accounts payable, distribution, etc?

☐ ***Set up your business***
- An accountant and/or lawyer can help set up the proper business entity.
- SCORE [www.score.org] is a national, nonprofit organization with 10,000+ volunteers who offer free, confidential business advice and mentoring from successful entrepreneurs & execs. They have resources for new and existing businesses, toolbox with templates, sample letters, business plans, etc.
- Most states have Small Business Development Centers to help with a business plan and get your business going. You can go online and check in your state and municipal government services directories.

☐ ***Learn about your industry***
- Go to your local stores that sell your type of product and learn who your competition is.
- Get to know what price points are acceptable in the low to high range
- Find out what materials are they using

- Find out what type of packaging/labeling is accepted by retailers
- Subscribe to that industry's publications.
- Attend trade shows
- The website [www.tsnn.com] has trade show schedules for every industry
- Get catalogs from the industry.
- Pay attention to the Brand and distribution company listed on the packages, as an example, General Manufacturing distributed by ABC Distributors Inc.
- Check out the product selection offered in specialty stores and mass retailers, low-end, mid tier and high end to see when your product would fit
- Keep an eye on the brands that are similar
  - You may be able to pitch to them as a potential licensing opportunity
  - Use them as a standard as you get ready to manufacture, to make sure will be accepted in the industry.

- *Find the right partners*
- To find engineers, manufacturers and their reps I use **www.thomasnet.com**
- For patent searches, companies like Patent Search International provide patent searches and opinion letters
- Connect with people at local inventors' groups and find out who they use
- Look on packages of similar products. Sometimes they have a distributor's name. This is known as a 'pre-qualified lead'. Call them up and pitch your product to them
- Licensing agents can introduce and pitch your product to companies they have pre-existing relationships with. The agent

takes care of the agreements and takes a percentage of your royalties.
- Stay clear of licensing agents that require upfront fees. Go with agents who will find the partner/partners to manufacture and distribute your product with very little risk to you.
- Just remember that the manufacturer has to have the retail distribution relationships to bring the product into the marketplace.

Do your research. I have plastic products so I subscribe to magazines like Plastic World that talk about plastic processes. However, plastic is not the only medium.
- You need to know what types of products there are that meet your needs
- How their processes work
- What is involved in the production of these materials
- How many different ways a prototype can be made

All of this research helped me when I called a prototype company and they told me how much the prototype would cost. When I asked what material they were going to make it in for that price, I took their information and called another place. They told me that for the same price they would make it in a different process using different material.

☐ *It can seem crazy at times*

It can be overwhelming when industry experts – designers, engineers and other professionals – explain things to you in words and phrases only they understand!

☐ *It's a time consuming process*

There are so many variables you have to be on top of all the time.

☐ *Your dream will keep you going*

So will the different people you get to speak with and the exciting decisions to make. When you see your invention developing and your business growing, your success and the success of your idea will be that much more rewarding.

Resources:

United Inventors Association **www.uiausa.com** has resources for inventors

CNBC has a show called The Big Idea with Donny Deutsche

Other network shows include American Inventors and Everyday Edisons

CHAPTER ELEVEN:
IF YOU DECIDE TO LICENSE

You can continue to make a living while you earn a percentage from the sales of your invention. Licensing allows you realize the potential of your product without having to expend the added time, money, commitment and risk.

- How do you feel about giving control of the success of your idea to others?
- Are you fine with making a small percentage of the profits?
- How do you know who to talk to?
    - ☐ The more people you speak to, the easier it gets. Talk to as many people as you can to get the background on everyone you may need to take your idea from concept stage to product on the shelf... from patent assistance, designer, engineer, setup/tooling, manufacturer, supplier, distributor, licensing rep and so on.

Rejection is tough. If you don't feel comfortable picking up the phone and making 'cold' calls, you can hire a licensing agent whose job it is to make these contacts.

As we said earlier, a licensing agent can be a tremendous asset, if for no other reason than the fact that their pre-existing status in the industry can sometimes walk your product through doors you can't access yourself.

Whoever they bring your product to is the licensee. Since the licensee is going to take care of the manufacturing and distribution, you need to know what their track record is, not only for quality of manufacturing but in their success as a distributor. There are publications that can help you get a better idea of the people you are planning on dealing with.

- License — a magazine for licensing all categories **www.licensemag.com**
- LIMA — International Licensing Industry Merchandisers' Association
- Global licensing information **www.licensing.org**

Stay clear of licensing agents that require upfront fees. Go with agents who will find the partner/partners to manufacture and distribute your product and take a percentage of your royalty fees as their payment instead. Licensing agents take a percentage, anywhere from 20% to 50% of the royalties. For best results, pick a licensing agent who has a track record with clients like you and/or products similar to yours.

Your royalties are not based on the retail price, but on the wholesale price that the retailer buys it for. The wholesale price the retailer pays includes what it costs to make it, what it costs to get it to the retailer (shipping, packaging, etc) and how much of a profit you want to make. The difference between retail and wholesale is this combination of what it costs the retailer and how much profit the retailer makes.

CHAPTER TWELVE:
HOW TO WORK AS A TEAM

Your team will be all the people you gather around you to help you realize your dream. It starts with your family. From there, you draw on your friends and fellow inventors for their insights and support. Your mentors will be team leaders in more ways than one, because they bring wisdom to the mix. Once you start transforming your big idea from concept to tangible form, you will start working with professionals whose expertise will help you reach your goal.

How do you work as a team?

As the innovative thinker whose big idea has transformed you into an entrepreneur, it is up to you to invite qualified professionals to join your team. You don't have to know everything they know: you have to learn to use what they know to enhance your success. You have to recognize their skills and reward their efforts in ways that matter to them.

Before you begin to build your team, you have to do your due diligence and educate yourself for success. We've talked about this before and now that you understand the process, you'll be prepared to do it again. Use all your resources to find the people best suited to join your team.

- You need people who have experience working with people and products like you and your big idea.
- You need people whose fees fit within your budget.
- You need people who are excited about your big idea and want to

be part of your team. The last thing you need is someone who is too busy or has to be persuaded to get involved.

- You need to talk to Small Business centers and use their resources to create a workable operating plan.
- Once you've done your research, you need to commit your energies to one plan or another – licensing route or manufacturing.
- Hire a patent lawyer.
- Talk to licensing agents
- Interview design engineers and get multiple quotes
- Research your product, its industry and its current niche in the market.
- Talk to an accountant and a lawyer to help you set up a business.
- Interview individuals whose administrative, management skills can help you.
- Check out various outsourcing agencies to hire freelance talent
  - ☐ www.Elance.com
  - ☐ www.Guru.com
  - ☐ www.coroflot.com
- Talk to manufacturers and get tooling and production quotes
- Talk to distributors, marketing specialists, printers, packaging companies
- Work with your mentor to build a committed team
- Invite them to give their input and act on their advice
- Recognize their efforts and reward them in ways that matter to them

Brian Fried

PART FOUR: THE ENTREPRENEUR

When their invention is actually out there, many inventors make the mistake of thinking that their job is done. In fact, their work has just begun. Like first time parents who feel a sense of accomplishment when their baby is born, you – and they – are suddenly realizing that a whole new set of responsibilities awaits you.

In this section we will show you what you need to do to keep tabs on your new 'baby', whether you have chosen to manufacture or license it to someone else. The next four chapters have been titled Farming, Fishing, Hunting and Nurturing.

<u>Farming</u> is all about gathering information about companies who you might want to work with and/or the way(s) that products like yours are being manufactured, packaged, promoted and sold. Your independent research involves visiting stores to see who carries what; what materials are being used; what kinds of packaging ideas currently exist; and what similar products retail for. You are also getting an idea of how these products are being showcased to the world.

<u>Fishing</u> is all about getting positive feedback from industry reviewers and satisfied customers. These third party references are an excellent way to verify claims made to you. At the same time, it is an essential first step as you build your network of support.

<u>Hunting</u> covers the tasks you must do to build new markets and find the market representatives that are best suited to help you increase your market share.

You And Your Big Ideas

Nurturing provides you with the tools to grow your data base of team players, colleagues and mentors.

Because there are two avenues you can choose from…

1. You can hire a contract manufacturer to simply manufacture your product while you stay in charge of marketing, sales and distribution, or

2. You can go in search of a licensee who run the whole show and pay you a royalty percentage based on sales.

…we thought it would be easier to compare the two models by creating separate columns and providing overviews of the work needed for each within the general categories of Farming, Fishing, Hunting and Nurturing.

CHAPTER THIRTEEN: FARMING

FARMING is all about gathering information on the ways an invention can be showcased to the world. FARMING is most often used when you are considering working with a licensee. That said, you can never have too much knowledge when you are starting something new, no matter which avenue you pursue.

As a licensor...	Or as the one calling the shots...
When you become a licensor, you give someone else the role and responsibility of taking your concept, manufacturing it to the highest possible standards and getting it to market. You no longer have full control and your input will be minimal because the licensee has their own way of doing things. You will always have the right to sign off on any final revisions, but in essence, you are giving them rights to use your idea or intellectual property. What you are expected to do is stay out from under foot and wait for the royalty checks to arrive in the mail.	With a vested interest in the success of your big idea, you owe it to yourself to take an active role in monitoring its progress. You need to increase your knowledge base at every opportunity so that you will be better informed and better equipped to handle the next big idea when it comes along. So where do you begin? You get a notebook and go shopping and/or go online to **www.thomasnet.com**. What are you shopping for? Knowledge! You're going to visit all the stores that carry products like yours and learn everything you can.

You And Your Big Ideas

As a licensor...	Or as the one calling the shots...
Ninety-nine times out of a hundred it will be in the licensee's best interest to do a great job when it comes to manufacturing, marketing, selling and distributing your product. They will have procedures in place that closely monitor the results. But no one is perfect. There can a time when even the best licensee fails to meet your expectations: • The company may be experiencing difficulties and fail to meet their deadlines and honor their commitments; • There may be marketplace issues that have come into play that affect the consumer's participation in the buy-and-sell equation; • The licensee may decide to prioritize another product.	You're going to go to the area in the store where your product would be found and write down all the company names that have similar products in their product line. There will often be a web address or a corporate address. If you have money in the budget, you should buy one or more of these similar products to help plan your sales pitch. Nothing impresses people more than someone who has done his homework and has a very clearly defined plan. Look at the packaging. Even if a product doesn't interest you, the way it has been packaged might. You may be able to incorporate these ideas to showcase your own product.

No matter which avenue you decide to follow... as a licensor or as the one calling the shots...

- You want to see how much competition there is.

    Can the market support so many versions of the same thing?

- Compare yours to the others on the shelf.

    Have you invented something so much better that it will blow the competition out of the water? If that is the case, you will want to use comparative strategies when you move on to the next steps.

When you've gathered the information, you will need to compile a summary report that allows you to retrieve the content when you need it. You should cross reference the report by product, by manufacturer and by store.

FARMING isn't done in a single day. You will want to get into the habit of visiting stores on a monthly basis. Buying trends are influenced by many external things: seasons, events, holidays and new innovations. Marketing campaigns and retailing strategies are adjusted accordingly. Products are moved and showcases are changed frequently to excite curiosity and invite people to buy.

[Author's Note: I also like to cross reference this information with quarterly reports from brand leaders that show trends and profits and losses from the products they offer. I take the names of these top companies and find out if they are publicly traded on the stock exchange. Then I look up the news and press releases written by the companies and review their quarterly results. You can even call the Investor Relations department of the company and ask them to send you the full quarterly results. You can learn quite a bit, not only about the company, but about your own future – imagining yourself running a company like this and seeing what is in store for you, both positive and negative, based on decisions made, the kinds of products and services that will make money and lose money – a day in the life of a big CEO running a corporation. One of the searching resources I use to do this is **www.yahoo.com**, but they all offer a similar capability. I go to the "Finance Section" and look up the companies I want to review and compare.]

CHAPTER FOURTEEN: FISHING

FISHING is all about getting feedback from industry reviewers and customers and building your network of support. It's about familiarizing yourself on product reviews, recalls, lawsuits and headlines. It's about putting all the information available into a context that allow you to stay one step ahead of everyone else.

As a licensor...	Or as the one calling the shots...
Knowledge is power. The more you know the smarter choices you are able to make. Just as the experienced angler knows where the biggest fish are, you will soon develop your own sources for valuable information that will leverage your smart choices and help you succeed. Your number one priority is your product. You need to ask for customer feedback to get a sense of your product's performance.	If you are manufacturing yourself... • Is there a flaw that can be corrected with a different tooling process? • Do you need to go back to the drawing board for a better design? • Would a different material be better, safer, simpler, more cost effective? • Has a natural evolution suggested itself and can a new product be launched? • Has some external force in the market place made the product in its current form unpopular, redundant and/or obsolete?

Your next priority is your team. You need to assess and review the work that will be performed on your behalf.

As a licensor...	Or as the one calling the shots...
• Does a prospective agent have a good track record? • What brands, products and/or services have they represented and how successful were they? • Is your licensee covering all the bases?	Once you hire them, keep on top of things: • Are the retail outlets doing all they could to display, advertise, promote your product(s) to the consumer?

Finally, you need to keep your finger on the pulse of the industry and the market your product serves.

- Read reviews on every product that falls into your niche... not just yours.
- Save the names and get to know the likes and dislikes of the editors that write about your industry and product in magazines, newspapers and technical journals.
- Save the catalogs that come to your house and use them as research tools. You'll be amazed by the way your perceptions change and you begin to notice packaging and marketing strategies more readily than you notice the products themselves.
- Subscribe to industry periodicals and publications.
- Attend trade shows
- Make a deliberate effort to introduce yourself to people who work in supply chains that may already be helping you.
- Join as many organizations as you can.
- Subscribe to their newsletters.

You And Your Big Ideas

- Attend their functions and ask lots of questions.
- Read books and contact the authors.
- Meet regularly with your mentor and actively seek advice.

Make notes of every conversation you have and compile a summary report. Cross reference your reports to make the retrieval process work for you.

CHAPTER FIFTEEN: HUNTING

Your FARMING and FISHING adventures will have netted you some incredibly valuable information, but this is only the beginning. You need to actively HUNT for ways to leverage the information you have acquired to build new markets and familiarize yourself with the individuals and organizations who will help you increase your market share. The easiest way to do this is to get on the Internet and start verifying the information you obtained by FARMING and FISHING.

- Search by
    - ☐ Product names,
    - ☐ Line names; and
    - ☐ Trade names.
- Look for articles and reviews.
- Look for headlines that announce:
    - ☐ Recent lawsuits filed and their outcomes;
    - ☐ New patents issued;
    - ☐ Trade shows.
- Look for websites and once you find them, go to their Contact pages to get phone numbers and the names of the contact people.

With all this information compiled you need to create an efficient retrieval system so that you can find the information quickly and easily. As your company grows, you will hire other people to do this

You And Your Big Ideas

administrative work. You won't be able to rely on your own memory to track down the person, place or thing you need. A data base will put names and related information into the hands of anyone who needs to access them. That means it's time to sort and record your information:

- By industry
- By service
- By function
- By product
- By person/contact

<u>With your data base growing, it's time to create a plan of action.</u>

You wouldn't believe how many people get excited about an idea and pick up the phone before they've decided on a plan. Ten seconds into the conversation they realize their mistake and find themselves wishing that the person they're talking to had been away from his desk or out of town because they have no idea what to say or do next.

You're probably tired of hearing this, but I can't say it often enough: The strategy that has the greatest chance of success <u>isn't</u> the one that promotes the idea that's the biggest or the best. It's the strategy that answers the question, "What's in it for THEM?"

What this means specifically is going to vary depending on the avenue you've chosen and the person you're talking to.

As a licensor...	Or as the one calling the shots...
A prospective licensee wants to know how much it will cost to manufacture the item and what price point it will sell for at retail. They will also need to decide whether it fits into their current or upcoming product lines.	A manufacturer wants to know how much work is involved, what the timeline is, how many will be produced and at what unit price. They will need to agree on material used, packaging, shipping arrangements and payment terms and methods.

As a prospective licensor, how do you translate these questions into a workable action plan?

1. The first thing you have to do is focus on your best product or your greatest strength.
2. Then you have to ask yourself what this product or service can do to enhance your chances of success.
3. Write down everything that comes to mind. Use lots of paper.

NOW

1. Do your best to stand in the shoes of the licensee and ask yourself what benefit <u>they</u> will get by focusing their time and effort on your product and you. It's not about your needs or dreams when their money is on the line.
2. Write down everything that comes to mind.
3. Do the same with everyone you plan to call. You're trying to overcome their objections and shape your message to get a Yes every time.

Now it's time to hit the phone! Or is it??? Do you know <u>why</u> are you calling?

Most people have no idea *why* they're calling. What they secretly hope is that they people they call will tell them. (I'm serious!) They say to themselves, *I'm calling a professional: an expert in the field. He'll know what I want. He'll tell me what to do.*

No he won't. If the busy professional even has the time to take your call, he or she will most probably say, "How can I help you?" and you'll be sitting there on the other end of the phone trying to think of an answer.

THAT's why you need an action plan.

Okay, you've already done your positioning homework, figuring out how other people will benefit from joining forces with you. Now you have to decide what you want them to do and why you want them to do it. Write down the answers to these questions:

You And Your Big Ideas

- What do you want?
- Why do you want it?
- What makes you think the person you're about to call can help you?

Before you make the call, check in with someone from your inner circle and tell him or her what you have in mind. Listen to their advice. When they give the thumbs up...

It's time to hit the phone! When you call, one of two things will happen:

1. The person you want to talk to will be ready, willing and able to talk to you.
2. They won't be there, or they'll be busy and you'll leave a message.

What one thing do you have to keep in mind as you open your mouth to speak?

What's in it for THEM! That's Right! Don't forget to be excited about your product and what it can do for THEM. Take a couple of long, slow deep breaths before you dial and smile as you speak.

Remember that you must be ready to accept rejection. If someone says No, it doesn't mean that you are a failure or your big idea is a dud. It means that this isn't the person who is going to take you to the top.

If you're not sure who to ask for when you place the call...

As the receptionist who directs incoming calls to either give you the name of person in charge of taking calls for **new product submissions** or tell you who is **in charge of licensing** for the company. Be sure to get them to spell the name for you and get their title as well. Ask for their office extension so that you have it on hand the next time you call. Then ask the receptionist to put you through.

1. Your telephone script if they CAN'T talk to you now and you're either speaking to their secretary or leaving a message on their voice mail:

My name is _____

I'm calling to speak to _____.

The reason for my call is that I have a product that may complement your product line because _____.

[Describe how it can do this. Keep it **under** 10 seconds.]

Continue for the secretary…

The potential market is _____. The estimated cost of production is _____. [No details at this point]. The suggested retail cost prediction is _____. [Again, no details]

[If you're speaking to a person] When is the best time to call again? Once again, my name is ____. I can be reached at _____. Thank you.

2. Your telephone script if they CAN talk to you now

My name is _____

I'm calling to speak to _____.
The reason for my call is that may complement your product line because it can save - or make - money [Tell them why or how. Do it in one sentence].

It's my understanding that you [summarize what you know about them]

I think my product could be of value to you [your company] because it [tell them very briefly what your product does and how it meets their needs]. I'd like to give you more information.

Do you have a few minutes now?

If they say yes, all you want to do right now is tell them how your product is a perfect fit for their needs. You want to use short, simple sentences and finish quickly by offering to email pictures of the prototype and descriptions from your design engineers along with a Non Disclosure Agreement. Confirm their email address and the correct spelling of their name and title. Thank them for their time. Repeat your name and say good-bye.

NOTE: They may not want to sign an NDA for any number of perfectly legitimate reasons. You will have to decide BEFORE you make the call that you are willing to share your idea with them at your own risk. If they don't want to sign an NDA, you have every right to ask why this is. You need to feel good about sharing before you share. If you don't like their reason – or the way they describe it – you need to be prepared to say 'Thanks for your time' and walk away.

Whether they invite you to send something or they decline, you DEFINITELY want to follow up with a thank you email. Why? Because emails create a paper trail. Be sure to put your name and the name of your idea in the subject line [for instance, Bob Smith and the Reversible All Weather Shoe] so they connect your name to the idea and have a greater likelihood of remembering you.

If you can't get someone you're speaking to on the phone to let you make a presentation that way, www.TSNN.com provides lists of trade shows for every industry, locally, nationally and worldwide. When you've found trade shows that showcase products that are similar to yours, try to make appointments to meet with the people who will be there and take your one-person show on the road.

As the one who is calling the shots... (You're calling a 'contract' manufacturer – someone you're going to pay to manufacture your product)

1. Your telephone script if they CAN'T talk to you now:

My name is _____. I'm calling to speak to _____.

The reason for my call is that I am developing a _____ product and require the services of a _____ provider. I have been told that this is your area of expertise and would like to arrange a convenient time to discuss the project.

[If you're speaking to a person] When is the best time to call again?

Once again, my name is ____. I can be reached at _____. Thank you.

2. Your telephone script if they CAN talk to you now

My name is _____. I'm calling to speak to _____.

The reason for my call is that I am developing a _____ product and require the services of a _____ provider. I have been told that this is your area of expertise and would like to arrange a convenient time to discuss the project.

It's my understanding that you [summarize what you know about them].

Would you be interested in meeting with me to discuss this in more detail?

[If they say yes] Would you be willing to sign a Non Disclosure Agreement?

If they say yes, get them to agree to meet with you. Confirm their email and other contact information and the correct spelling of their name. Thank them for their time. Repeat your name and say good-bye. Whether they agree to meet you or they decline, you DEFINITELY want to follow up with a thank you email. Emails create a paper trail.

CHAPTER SIXTEEN: NURTURING

NURTURING provides you with the tools to grow your data base of team players, colleagues and mentors. NURTURING is all about managing your relationships.

Relationships are founded on a give-and-take equation. As one half of the relationship you forge with another person, the most powerful qualities you can bring to the table are:

- Enthusiasm
- Commitment
- Consistent Action
- Unconditional Support

Once you've built up a relationship with a licensee or a contract manufacturer directly, nurturing is all about taking care of that relationship:

- Providing what they need when they need it
- Maintaining good relations with your contact people
- Educating yourself to be able to respond to their needs
- Thanking them for all that they do for you
- Staying on top of things – establish consistent Q&A reviews

[NOTE: As a licensor, when you get your quarterly royalty report and payment, this is a good time to conduct an overall review and evaluation.]

Nurturing is about thinking of new ways to expand your product line and build on the relationships you already have. Always tell your licensee or your contract manufacturer that you are willingly giving them the lead. At the same time, make sure they know that you intend to take an active role and that you are keen to be as involved as you can.

Nurturing works in another way as well. Just as licensees will promote and sell products that come from many sources, you will begin to find new and different avenues that work for you. You may be ready to take the reigns and call the shots. You may cross paths with other licensees who may be interested in different product designs and concepts you have to offer.

If you are a licensor and have a licensing agent, he or she will be working for you. As your provider, they have a responsibility to you. As their client, you have responsibilities too:

- You need to pat them on the back for the work that they do;
- You need to read their royalty and review Sell Throughs or retail sales reports your agent has obtained from the licensees they have signed up for you, which provide statistical tracking on your product(s) by date, by location, by color, by event, by store… and more, and make them aware that you are excited by the progress and eager to help them achieve more;
- You need to make it clear that you are committed to their success because when they succeed, so do you.

As you continue to do your research, you will meet more people and discover new ways to expand your market share. You will also find yourself coming up with new product concepts and improvements on the old product concepts. In this way you are nurturing the relationship you have with your company and your long term success as an inventor.

Always be open to new relationships. The more people you know who enter into relationships with you, the stronger your team and the greater your success.

As long as you honor your commitments and do not work at cross purposes, undermining the success of the relationships you already have, this kind of nurturing is an essential step in growing your business exponentially, no matter which way you choose to run your business.

PART FIVE: LIVING THE DREAM

Congratulations and welcome to the life you always dreamed of living someday!

Now that 'someday' is here, it's time to give you an overview of the work you will need to do to keep your Inventing Empire alive and well, supporting its inventions and providing everyone who becomes involved in your business with a clear set of rules and operating procedures they can understand and follow.

As a business owner with many inventions and ideas in various stages of design, development and completion, you will quickly outgrow the hands-on approach you used in the farming-fishing-hunting-nurturing model you learned in Part Four.

At this stage of the game, you will be in need of more advanced techniques and tools and we have provided them in the four chapters that follow.

CHAPTER SEVENTEEN: PACKAGING

What is a package?

A package serves two purposes, only one of which has to do with keeping something safe and clean. A package also provides unlimited opportunities to showcase and sell whatever it contains.

What packaging means to you

As an inventor and an entrepreneur, you are starting to realize that you have to learn many new skills. You need these skills so that you can achieve excellence in everything you do.

- As a businessperson
- As a pro-active team player
- As a professional working among professionals
- As an inventor

The first package you need to design and produce is the one you use to present yourself to the world.

- As a businessman, you need to present a businesslike image
  - To command respect you must dress for success
    - o You don't have to spend all your money on your wardrobe, but you do need to wear suitable clothes.
    - o Talk to your mentor and take note of speakers and other professionals.
  - To indicate quality you must use quality presentation tools
    - o Don't skimp on cards and letterhead.
    - o Get a good portfolio and package your materials professionally

- ☐ To underscore commitment you must establish a presence
  - o You need a computer, printer and scanner
  - o You need an office with a desk and a filing system
  - o You need a website
  - o You need a dedicated business phone
  - o You need a voice mail retrieval system
  - o You need an email address
- As a pro-active team player your package must be user-friendly
  - ☐ Packaging is how you represent yourself to the people around you. People learn who you are by the things you say and do.
    - o Keep your promises
    - o Honor your commitments
    - o Focus on the What's in it for THEM equation at all times.
    - o Listen and give thanks
- As a professional working among professionals your package must meet the highest standards and pass the most rigorous tests:
  - ☐ Everyone faces challenges in life. Those who take on the responsibility that comes with being a professional take ownership of their lives in a unique and special way. As a professional you will be expected to:
    - o Accept nothing less than your best from yourself
    - o Inspire those around you to achieve excellence for themselves
    - o Lead by example
    - o Recognize effort and reward results
    - o Bring joy into the lives of others
    - o Share the wealth
- As an inventor you must fill your package with the ingredients that are essential in the recipe for success
  - ☐ Dare to dream
  - ☐ Push the envelope

- ☐ Believe in yourself
- ☐ Embrace challenge
- ☐ Invite change
- ☐ Welcome insights
- ☐ Build a team
- ☐ Hunger for knowledge
- ☐ Live life to the fullest
- ☐ Give thanks for your gift

The other package you need is the one that showcases your product. Here you can turn to the experts to help you design the package that does the job right.

- Work with the contract manufacturer or licensee to confirm that your packaging conforms to the rules that the retailers require.

  [NOTE: Remember that a contract manufacturer is a producer of products.

  He has no distribution channels and will not know what the retailers require.

  The licensee should know, because distribution is part of the work they do,

  but it's always best to double check to be safe.]

- If you're going to be packaging on your own, ask the manufacturer what kind of packaging they recommend as well.

- Some manufacturers only make; other people package. Find a packaging design company and work with a graphic artist to create a concept.

- Websites like Elance.com, Guru.com and Coroflot.com offer outsourcing facilities where you can hire someone to do the work you require.

- Packaging design is also about the information that is on the package. Make sure that the copy clearly describes the product and what it does. This includes various uses, how it is best used and who will use it.

- Does your product stand out by itself or does it need some truly inventive packaging to make it stand out on the shelf?

You want to have the best packaging you can afford for your product. In the beginning, you may have to cut a few corners. Once you get some revenue coming in, you can think about making your design better. One consultant who gives advice to inventors is **www.PackagingDiva.com**. It's worth taking the time to check out this website, among others, and get some fresh, inspiring ideas.

CHAPTER EIGHTEEN: PROMOTING

When we think of promoters and promoting, the vision of a slick huckster with a glib tongue floats in front of our eyes. The word brings with it a taint of dishonesty that makes many of us feel uncomfortable, doesn't it.

The reason we get a negative feeling is that we've experienced disappointment when we've invested our belief in something that failed to measure up to our expectations.

Is it always the fault of the promoter? Not always. Many times the fault lies with us, for investing our happiness in something that exists outside ourselves.

As you get ready to promote yourself, your products and your business to the world, you need to take a moment and do a bit of soul-searching. You have to take a realistic look at the package you are creating and the strategy you are building and ask yourself:

- Am I telling the truth?
- Am I delivering a value-added product or service?
- Am I building a reputation that will stand the test of time?
- Am I laying the foundation for long term success?

If you can honestly say 'Yes' then you're half way there before you've begun.

The tools a promoter uses to excite the public about a concept appeal to the five senses in order to capture the imagination and stimulate a response that promotes a commitment to action:

- Sight
    - ☐ Images and words inform. Knowledge is power.
- Sound
    - ☐ Sounds trigger action. Action inspires change.
- Smell
    - ☐ Smells bring back memories. Memories build confidence.
- Taste
    - ☐ Tastes satisfy cravings. Satisfaction elevates mood.
- Touch
    - ☐ Tactile qualities access comfort zones. Comfort stimulates commitment.

When the package is put together and the message is sent to the brain, the hoped-for response is one that elevates this product above others in the mind of the buyer. The 'What's in it for THEM' question has been asked and answered.

The result is success.

As you go in search of the best methods to promote your product, your business and yourself, with the all important truths taken care of, all that is left to decide is how you want to tackle the challenge.

Whatever path you choose, you need to get input from the experts who will help you promote your idea from your initial proto-type and give it the foundation it needs for a long and productive life.

- You will need to set up appointments with local providers, potential licensees, licensing agents, design engineers and contract manufacturers.
    - ☐ What is the message you want them to send to the world?
- Set up other appointments with out of state companies at tradeshows

You And Your Big Ideas

- ☐ How will national representation reinforce your success?
- Have your prototype, any patent protection, trademarks, business plan including research and projections if possible prepared for your presentations.
  - ☐ What are the key elements you want your presentation to promote?
- Do your due diligence before selecting people and companies to work with.
  - ☐ What qualities must they bring to the table?
- Prepare your cost estimates and double check your budget figures
  - ☐ What value can logic, patience and commitment add to the mix?

CHAPTER NINETEEN: PITCHING

'If you build it, they will come.'

That's what the voice told Kevin Costner in the film Field of Dreams.

It's a lovely thought, but unfortunately they didn't complete the sentence in the movie and as a result, people went away thinking that people will automatically invest in something once it has been built. This isn't true. What they should have said, to make it real, was this: If you build excitement, they will come.

Building excitement is what pitching is all about. When you share your enthusiasm and capture the imagination, curiosity does the rest. How you generate curiosity and capture the imagination is by framing your statements to promote your idea as the answer to the question, "What's in it for them?"

Let's look at a few examples to help you see pitching at work. It will be up to you to take these examples and apply the ones that are most suited to your needs.

A couple of years ago, I invented the *Balloon-o-Band*: a nylon wristband with a metal D ring and Velcro at the ends of the band for easy on and off. This product allows children to 'wear' their balloons when they go to the circus or the county fair. I have a young child of my own and after one too many days spent holding my daughter's balloon and/or listening to her tears when she accidentally let go of the string and the balloon floated away, I suddenly thought of a simple solution that would save everyone a lot of heartache. When my daughter wants to take the balloon off between rides at an amusement park, she can. If she lets go, the band also serves as a weight so the balloon will not fly away!

How did I pitch this product? I spoke to the child in all of us. I spoke to the parent we have all become. I spoke to the balloon seller who has to compete for his business in the hot sun. I spoke to the manufacturer who – up until that moment – thought he'd heard everything about balloons and balloon accessories.

Was I successful? You bet!

I have a friend named Roger Browner who came up with the concept of round cheese after he got tired of the way square cheese slices kept dripping off the sides of his burgers. It didn't make sense to him to have a round burger with square cheese! He went through many trials and tribulations to get cheese produced in round slices. Then he decided that he needed something more to get past the buyer judges, so he added flavors: smoked hickory, sharp cheddar and Habanero Jack. With a new look and specialty flavors, he was ready to pitch *Gourmelts Round Gourmet Cheese*: A different product in the dairy section and a way to solve a problem for the BBQ season!

A product that I am having success with – it has been accepted by a licensee and being shown to mass and specialty retailers – is a space-saving idea that everyone likes. Here's the pitch…

When you buy a carton of eggs and put it in the refrigerator, it takes up the same space whether it holds twelve eggs, six, three or just one. My solution is called *Eggstra-Space*: A collapsible egg tray designed to maximize storage space in a unique design that protects a popular food product common in most refrigerators.

A member of The Inventors and Entrepreneurs Club of Suffolk County came up with the *Game Chamber*: A revolving mechanism for the small game cartridges for Nintendo DS games. Here's the pitch…

Kids, here's an easy way to keep your games tidy: Now you have a storage unit that won't let you take out another game until the one you're using goes back in! The product is available in some mass retailers, a few dot com websites and he's currently working on expanding his distribution channels.

CHAPTER TWENTY:
THE PERPETUAL GROWTH MACHINE

What is the perpetual growth machine? It's you!

When you take the information we've provided on the preceding pages and put it all together in a manner that works for you, there'll be no stopping you!

What this chapter is meant to do is review the information and reinforce the message we have presented to you. We've structured it as a checklist that you will keep with you... when you talk to your mentor, when you attend your group meetings, when you invite the professionals to join your team. It is an essential tool, this checklist. We can't urge you strongly enough to use it every day.

Your Checklist for Success

1. Am I an inventor? Am I patient enough? Do I have unconditional belief?

2. Am I committed - and not just interested - to make this happen?

3. Do I have Attitude? Do I embrace the challenge that comes with change?

4. Have I done my due diligence? Have I educated myself for success?

5. Am I a team player? Do I nurture those who nurture me?

6. Am I professional? Do I achieve excellence in all that I do?

7. Do I understand the processes?

You And Your Big Ideas

8. Am I asking the right questions?
9. Am I listening to the answers?
10. Do I know when to let the experts do what they do best?
11. Am I able to resist trying to reinvent the wheel to satisfy my own ego and welcome the knowledge that the experts can bring to help me overcome my challenges?
12. Am I doing all I can to follow through and support my big idea?
13. Am I using a business model that enhances my success?
14. Do I remember to ask: What's in it for them?

CONCLUSION

SHOOT FOR THE MOON and REACH THE STARS

The dictionary is the only place where success comes before work.
- *Vince Lombardi*

Ambitious dreams inspire greatness and make you happy; yet the world is full of people who would rather not reach for their dreams in case they fail. One thing is certain… if you don't try, you will never know what the outcome might have been. You will never know what it's like to shoot for the moon and reach the stars.

A lot of parents teach their children to be modest in their expectations. They don't want to see their sons and daughters hurt so they teach them to have ambitions that are well within their capabilities. Hoping to help them avoid the pain of failure or the embarrassment of having other people know they've failed, they doom their children to a life of predictable mediocrity.

For the true entrepreneur, 'getting by' is not a goal that makes you glad to get up in the morning. Doing a job you don't enjoy because it pays the bills is not a definition of success or happiness. Suppressing your ambitions, talents and dreams is a recipe for unendurable frustration that gets worse over time, not better.

Your mixture of interests, talents, dreams and inspirations is unique. It has never appeared in human history before and it will never come again. This means that YOU are capable of achieving something truly remarkable – something that no one else has ever done before – and if YOU don't go for it, no one else ever will.

You And Your Big Ideas

It might be years or even centuries before someone else stumbles on the same discoveries you have made and writes the books, sings the songs or sells the products that are within your grasp. Until that happens – IF that ever happens – the world will have to do without them.

Let's take another look at Thomas Edison: the man who tried 10,000+ times until he perfected the light bulb. He had a total of three months' formal education but that didn't stop him from following his dream. When most people laughed at him he kept trying. He had to invent the light bulb because it would pave the way for all the dreams he had after that. Once he had his light-bulb, Edison knew he could change the world.

Eventually, his efforts were rewarded. With his light-bulb in production he went on to apply for and receive over 1,000 patents — more than anyone in history. To this day his record remains unbroken: Thanks to a dream that wouldn't die, the man with three month's education became one of the richest men in America.

Edison had ambitious dreams that inspired and energized him and fuelled his commitment when the road was hard and full of setbacks. If you are willing to believe in your dreams and commit yourself to see them through, they can and will come true.

Make Your Ambitious Dream Come True

- Choose a dream that inspires you no matter how hard the going gets.
- Write down exactly what you're aiming to achieve.
- Explain how it can help others and change the world. The best way to get the things you want Is to make sure other people get what THEY want, first!
- List the steps you need to take to make it happen.
- Work on those steps each day and watch the magic start to happen.
- Keep your outlook absolutely positive.
- Learn from your mistakes.
- Adjust your plans accordingly
- See yourself on your way to great success.

When I set out on my journey to write this book, I wanted to create a portable mentoring guide that identified the basic tools innovative thinkers, inventors and entrepreneurs must acquire to achieve their goals.

As I had learned, it's not enough to tell a person *that* they must do something... what I needed when I was starting out was a way to understand *how* those things were done. I needed to know what each of the individual steps were, when to apply them and why they had such a profound affect on the end result.

When I felt I had everything organized into a model that had the best chance of helping others I sat down to write this book, knowing that the "don't shoot the messenger" adage exists for a reason. Even with the best of intentions, it isn't possible to provide an absolutely foolproof model that guarantees 100% success.

What I mean by this is that the last thing I want you to do is pin your ultimate hopes on anyone but you. At the end of the day, your success does not belong to me, to your mentor or this book. It belongs to you.

I've done my best to create a reference guide and mentoring tool that will help you reach that goal. I've included information and advice designed to educate, inform and warn you to the full extent that I can. But as everyone knows, the only things certain in life are Death and taxes. Even the best idea sometimes has to wait for its time to come.

My role is to encourage you in your efforts by educating you to the realities that exist in the inventor's world. I'm not trying to give you slick tricks that promise success... I'm trying to ground you in solid principles that will sustain you in good times and bad. I've made a concerted effort to deliver a product that will continue to enlighten and inform you as you take each step toward your goal.

Remember that you are not traveling this road alone. All around you, there is a team of well-wishers and supporters ready and waiting to help you. When the going gets tough, go back through this book, talk to your mentor and turn to the Resource Guide that follows.

Good luck and Think UP!

BEST BETS

We've referenced several organizations, agencies and individuals on the preceding pages whose contributions to the inventing process make the road much, much easier to follow. We've also provided their contact information in the reference guide that follows.

However, we felt this portable mentoring guide would be incomplete if we didn't take the time to recognize those individuals who stand apart from rest and give you a more detailed description of the work they can and probably will do for you.

Remember that these people are hard-working, busy professionals. If/when you contact them, be patient: they will get in touch with you as soon as their prior commitments allow them to.

Without further ado, please say hello to Cheryl Perkins, Don Debelak, Jack Lander and Richard Klar...

Cheryl Perkins: Innovation Edge

CHERYL PERKINS is... a thought leader in innovation and a creative catalyst in brand-building initiatives for companies looking for that innovative edge.

In 2006, Business Week magazine chose Cheryl as one of the Top 25 Champions of Innovation in the world. She was also named as a top executive driving vision within the consumer goods industry (Visionaries 2006) by Consumer Goods Technology magazine.

With over 20 years experience directing growth and innovation as the Senior Vice President and Chief Innovation Officer for Kimberly-Clark, Cheryl ran the company's innovation and enterprise growth organizations, including research and development, engineering, design, new business, global strategic alliances, environment, safety and regulatory affairs, and oversaw innovation processes, systems and tools. She has ten U.S. Patents and several more pending.

Cheryl brings her expertise to **Innovation**edge, creating new strategic business opportunities for companies seeking a competitive advantage to ensure long-term growth and deliver a continuum of sustainable innovative solutions. Few consultants are able to identify and transform insights, designs, technologies and capabilities into total solutions and new-to-the-world innovations as Cheryl does.

Contact information:
www.innovationedge.com
Email: cperkins@innovationedge.com Phone: (920) 967-0470
1526 S. Commercial Street, Suite 200, Neenah, WI 54956

Capture the WOW

From large corporations to startups, innovators are looking to be recognized as forward thinking in their ability to design, develop and deliver breakthrough innovation in a way that says "wow" to consumers. Innovation is about creating and delivering new and more effective products and serves to consumers better than anyone else. A successful innovation creates value to society, along with a "wow" factor—that emotion and passion state consumers enters when they see a new category that completely amazes them.

To capture that feeling, corporations are looking for products based on discovering insights into how shoppers and users think and feel about the world in which they live, work and play. That's why so many companies and inventors must study and leverage trends, global events, economic forecasts, politics, war and other things that impact the shopper's values, fears, desires and spending habits. Those trends might include a continued movement of manufacturing jobs offshore, rising health care, food and fuel costs, high-tech gadgetry that keeps us connected and entertained, online social networking, the credit crunch and the changing housing market.

Innovative companies study these trends and learn how to leverage them to provide even more solutions to their customers. Inventors and entrepreneurs recognize that a true innovation affects people by changing their habits through new industries or categories.

Delivering that wow-factor to consumers means that companies must look beyond their own brick and mortar facilities to find the talent and capabilities they need to grow. I'm seeing more and more companies and entrepreneurs embrace the idea of Open Innovation in which they choose strategic outside partners and inventors to deliver new products to the marketplace.

The most exciting dynamic happens when businesses discover partners with just the right mix of unique and complimentary capabilities. It's no longer a competitive advantage to find these partnerships, inventors and networks, it is a competitive *necessity*. That's good news for inventors, and it's the reason Open Innovation models are so important for businesses looking to give their consumers something of value. They need fresh new ideas and new ways to use them in order to create a competitive advantage and deliver sustainable, innovation-driven growth and profitability.

Just as many corporations are challenged when it comes to developing sound strategies to overcome innovation challenges such as the "not invented here" syndrome, inventors also face barriers to getting their products to market, and finding that right fit with companies that could take their ideas to a new level. Getting an invention to the market place is like connecting an electrical circuit with multiple components that all need to be in place, in the right order, for the circuit to function and the energy to flow and do useful work. This circuit depends on the connection to the partners to get the invention plugged in to the market.

Inventors who want to set themselves apart will do well to think outside the box of they way they approach businesses to partner with.

Not only must they be focused on their invention, they must also convince businesses that they have their finger on the opportunities for a series of solutions for unmet needs in the marketplace. It's critical that inventors recognize that most external partners are more interested in building a pipeline rather than in just finding a single new product. Continuous innovation allows leaders to deliver growth over time.

Once inventors understand that true innovation must be a sustainable, repeatable process that helps companies adapt to constantly-changing conditions and trends, they can position themselves as valuable partners who will help deliver solutions that will make their customer's lives better.

So how do inventors and entrepreneurs fill that need? Typically inventors are focused on their idea or invention, and may miss the opportunities for a series of solutions for unmet needs in the marketplace. That's why pipeline planning must be a critical part of the business plan. Inventors need to recognize that most external partners are more interested in building a pipeline rather than in just finding a single new product.

Patents are usually first on the priority list for inventors, along with obtaining copyrights, trademarks, industrial design rights and trade secret filings.

Building a strong Intellectual Asset Strategy and Asset Portfolio is also important to maintain a prolonged competitive advantage with a product or service. Inventors are often surprised to find that just filing a patent is not enough to secure competitive advantage.

Intellectual asset (IA) refers to creations of inventions including copyrights, trademarks, patents, and related rights. Inventors have exclusive rights to the invention which is covered by their IA. Patents, trademarks, and designs rights are sometimes collectively known as **industrial property**, as they are typically created and used for industrial or commercial purposes.

Within their intellectual property, inventors must also understand different licensing components such as term, which is valid for a particular length of time, and territory, which may limit you to a certain region such as Japan or the U.S.

Brand licensing is the process of creating and managing contracts between the owner of a brand and a company wanting to use the brand in association with a product, for an agreed period of time, within an agreed territory.

Brand licensing is well-established business, both in the area of patents and trademarks. Mickey Mouse's popularity in the 1930s and 1940s continues to result in an explosion of toys, books, and consumer products with the lovable rodent's likeness on them, none of which were manufactured by the Walt Disney Company.

Brand extensions makes the brand licensing marketplace much more lucrative, as companies rent out their equity to manufacturers. Instead of spending untold millions to create a new brand, companies were willing to pay a royalty on net sales of their products to *rent* the product of an established brand name. So successful has the process become that some companies like Harley-Davidson and Nathan's make more money from licensing than from manufacturing!

What this all means for inventors is that they need a broad array of intellectual assets solutions to stay competitive. One of the best tools for success is an Innovation Roadmap. The roadmap should include a gap analysis between the inventor's innovation and intellectual asset systems with corporate business plans, and identify important gaps between objectives and beliefs versus performance. This deep-dive analysis should include leadership skills, systems and tools, communication, success in realizing value from intellectual assets, and other corporate strengths and weaknesses.

Many corporations focus their intellectual asset efforts on pursuit of traditional patents for materials, products, and methods of manufacturing. However, some of the most important innovations in many companies reside in other areas and may be left unnoticed and unprotected by those guiding patenting efforts.

There is an increasing need for companies to recognize and protect innovations in "business method" areas such as logistics, ERP systems, marketing research techniques, telecommunications, business models, distribution channels, process control and quality systems and data mining strategies.

In addition to patents, inventors can and should protect their business method innovations with a variety of other approaches, such as publications, strategic partnerships with contractual agreements, digital intellectual assets, and other tools to help protect those unique innovations.

Don Debelak

DON DEBELAK is... the author of four of the best-known invention books of the last 15 years. He has been working with new products and inventions for over 25 years and was the author or Entrepreneur Magazine's Bright Idea column on inventions for over seven years. Don has spent his career marketing products for new and small businesses, writing numerous business plans for raising money, both from investors and banks. Don has worked with all types of business, especially as a consultant for the University of St. Thomas Small Business Center, from small one man service business to high-tech ventures that are set up to raise money and launch a new product.

Contact Information:

www.dondebelak.com

P.O. Box 120861. New Brighton, MN 55112

Phone: 612-414-4118   Fax: 651-773-5866

Outsource Entrepreneur

Inventors are mostly familiar with licensing an idea, which is where another company acquires the rights from the inventor and pays a royalty. License deals can be hard to put together because the company licensing the idea has risk involved and they may want to see if the inventor can succeed with the product on their own. An option to licensing is for an inventor to be an outsourced entrepreneur, working with a manufacturing partner and a marketing partner to make and sell the product with low risk.

There are essentially three aspects to bringing a product to market: research and development; manufacturing; and marketing. As the inventor, you perform the task of research and development (inventing and developing the idea) and outsource the manufacturing and marketing to investing partners who are willing to invest their own money in the project in return for either an exclusive manufacturing agreement or marketing agreement. The investing partners provide further development and pay for many of the start up costs. In return they will receive more profits than contracted work and will have some control of the idea. This is often the fastest way to bring a product to market, is low risk and allows you to move onto new ideas quickly since the other partners will continue to manufacture and market your idea without excessive involvement on your part. This is not for everyone and will require you to have strong deal-making skills, but the low-risk and low-investment aspect makes this appealing for most inventors, who are often strapped for cash.

Finding Potential Partners

Not every manufacturer or marketer will be a good candidate for outsourcing. You need to find manufacturers who can add your product without too much investment in new machinery and who are running below capacity. Marketing partners must carry products similar to yours and your product must represent a significant increase in sales for the firm, at least 20% so the marketing partner has real incentive for taking on a new product.

To find a manufacturer, start with your state industrial directory (available at larger libraries) and look for manufacturers who could make your product. If your state doesn't have the right type of manufacturer, look in the Thomas Register, either on line or at your library. You can also get help from SCORE, www.score.org, which is a group of retired executives that help small businesses. Once you have the names of manufacturers, see if they would want to do contract manufacturing for you. If they say yes and are aggressive about pursuing your business, they will be a candidate to make the part and wait for payment until you get paid from the sale of the product in return for an exclusive manufacturing agreement.

To find a marketing partner, look for companies that sell the product to the same distribution channel and end user you want to sell to. For example, if you want to sell bike accessories to specialty bike shops, you would look for other companies selling bike accessories to bike shops. You would approach those companies to see if they wanted to private label your product or sell it under an exclusive sales agreement. A private label agreement is where you would put the company's name on your product and then the company would sell it as a standard part of their product line. An exclusive agreement is one where you would only sell the product to that one company. If the potential marketer is interested in either, you have a strong candidate to be your marketing partner. You can also get help from SCORE or the Small Business Development Center near you.

Working the angles

Approaching your partners can be difficult. Manufacturers don't want to sign a deal unless there is a marketing firm guaranteeing sales and marketers don't want to sign a deal unless a manufacturer can promise to produce the projected volumes at a reasonable quality. The best way to put the whole deal together is by first approaching a manufacturer and ask them to commit to the project if a marketing firm will sign on to the deal. If the manufacturer agrees, then you have a solid enough commitment for a marketing firm to be willing to sign on. Both parties need to know the other party is involved in the project in order to move forward.

Summary

In many cases outsource entrepreneurs make 10 to 12% profit, versus a 3 to 6% royalty when licensing, and they have much more control over their product's success and the direction the company follows. But it requires you to be a firm negotiator with high sales and marketing skills to convince companies they will gain by joining in with you and the other partner. If you have an idea with strong potential where the costs and risks for manufacturers and marketers is low, you are a good candidate for becoming an outsource entrepreneur.

Licensing

Licensing is what happens when a company takes over your new product idea and pays you a royalty of the sales from your idea. The licensee can be a manufacturer, marketer or a product development company. Since licensees take on all the risk of a product, they are cautious about what products they will license. Most companies will only license an idea if they are fairly certain it will be successful, so it is up to you to convince them. While the earning potential is lower than the outsource entrepreneurial approach, many inventors choose this strategy because once you license the idea you have no more responsibility.

Finding Licensing Candidates

Many inventors fail at licensing because they choose the wrong companies to approach. Many people want to license to the big companies without realizing that these companies rarely, if ever, license ideas from inventors. Unless you have a truly innovative product, you should not approach a company with a major market position. Fortunately, companies with smaller market shares are always looking for ways to expand their share and break into new markets and are therefore more willing to consider licensing. Since a licensed product produces a lower profit margin due to the royalties paid to you, you need to find companies that would benefit greatly from adding your product to their line.

There are two important parts to finding the right licensing candidate, first finding the right companies and then finding the right person at the company who can help you push your product through to a licensing deal.

The best way to find candidates is to attend industry trade shows where most if not all potential licensing prospects will be attending. Look for companies who sell to the same end user and through the same distribution channel. Start by looking for companies that are trying for the same sort of solution, but where your product will add value. For example, if you have a new better way to clean a deck, ideal companies would be ones that offer products to paint the deck, or companies that offer products to clean siding. Your product will help those companies provide a better solution to the customers' need for a clean outside appearance of their home. A second choice would be companies that offer cleaning products for the outside of homes and a third choice would be companies that sell general cleaning products.

While at the show get the business cards of sales people, independent sales representatives and marketing people at your target company. Show those people your product, either at the show or later on, and ask them if they would be interested in helping you present your product to the company. Normally, if they like your idea, they will help you because they will create a positive image inside their company for being a "go-getter" and bringing in promising ideas. If you can't go to a trade show, start getting trade magazines and start sending off for literature from any company with a product line that fits in with your idea. The literature often comes with a salesperson's card: this is someone you can use as a helper when you are ready to try and license your idea.

Deciding on Your Terms

Before you approach your licensing candidates, you want to be clear on what terms your idea will be licensed and what percentage of sales you want to receive. You can license your idea to just one company, which is an exclusive license or to many companies, which is a non-exclusive license. You will also want to determine what exactly you are licensing: either the product or the technology. Be clear on what you want but still be willing to negotiate, but remember to protect your future rights to the idea. For instance, if you license just your product, any advancements will exclude you from royalties; but licensing the technology could have you receiving royalties for many years. You should expect a royalty of between 3 and 7% if you license an idea.

Prepare a Licensing Presentation

When you find interested companies you will be asked to come to the company's office for a presentation. This presentation should not last more than 15 minutes and you should allow for questions after the presentation. If at all possible, you should include a demonstration. This is the most effective way of selling your idea. If you cannot provide a demonstration, try to incorporate a five-minute video showing people using your product. During the remainder of your presentation, you need to show the company why the product will be successful and that your product matches the company's goals and current market strategies.

Jack Lander

JACK LANDER is... is a mentor to inventors. He has written a popular featured column for *Inventors' Digest* magazine for the past eight years and served as the invention development and manufacturing expert for *Entrepreneur.com*, the Internet arm of Entrepreneur magazine. He has also written and published a book on job searching, and was commissioned to write *Make Money by Moonlighting* by Ted Nicholas of Entrepreneur Publishing Co. He produced and edited *THE Inventor's Master Plan* for the United Inventors Association, and authored the chapter on prototyping for Don Debelak's book, *Think Big*. He has also written *How to Finance your Invention or Great Idea*.

Jack has served as President of the prestigious United Inventors Association, a not-for-profit umbrella organization that helps inventor networking groups and inventors throughout North America. He presently serves his ninth year as Vice President of the (not-for-profit) Yankee Invention Exposition and Yankee Entrepreneur Workshops, held each October in Waterbury, Connecticut. He founded Innovators Network, a local inventor group, and The Inventor's Bookstore, a "dot-com" business that is now a not-for-profit subsidiary of the United Inventors Association.

Jack is a mechanical engineer who has received ten patents on highly successful laparoscopic surgical instruments, computer chip testing devices, and a bicycle transmission that is more efficient than the traditional derailleur. He was the founder of Shortrun/Precision Fabricators, a business specializing in producing prototypes and short production runs for high-tech businesses in the Los Angeles area. He has helped more than two thousand inventors in the U.S. and other countries and coaches inventors in all aspects of the invention process.

Contact information:
www.inventor-mentor.com
Email: Jack@inventor-mentor.com   Phone: 203-264-1130
U.S. mail: 949A Heritage Village, Southbury, Connecticut 06488
USA

Marketing Your Invention

Marketing your invention means either licensing it for royalties (or an outright sale), or producing it and selling it. Most inventors prefer to license because they don't have the time or resources to produce and sell. My emphasis herein will be on licensing, although all of the principles apply to producing, as well. The licensing approach can be summed up in four points:

1. Connect
2. Seduce
3. Negotiate
4. Sign

Connecting means more than simple contacting. It means that you are acknowledged — that you have had a response. Sending off a letter to an unnamed marketing director is not connecting; it is contacting—in fact, it is merely *attempting* to contact. You have no assurance that your intended recipient has received your letter until you get a reply.

Contact and response with a named and titled person, either in-person, by telephone, by e-mail, or through that person's assistant, is connecting. Remember the salesman who bragged that he had gotten two orders on a certain day? The orders were, "Get out and stay out!" Not a very good day. But at least he had connected, and had no doubts about the intentions of his prospect.

All corporations of any size or sophistication will not review inventions until you sign away all rights except those granted by your patent. Should you do so? You can't get your foot in the door unless you do.

Seducing means not merely showing and telling about your invention, but arousing sufficient interest that your connection is receptive to knowing more—perhaps even to being agreeable to negotiations. Seduction, as I use it here, is not deception or the taking of unfair advantage. It is the preparation of the initial paper sales pitch in a manner that is honest and compelling, but which greatly excels the ordinary approach.

Negotiations don't promise that you'll have a deal. Many negotiations fall apart after the corporate lawyers get involved, and write outrageous terms designed to give their clients (or employers) one-sided advantages. Sign at your own peril if you don't have an intellectual property attorney (a patent lawyer) review the agreement first. Better yet, go into any negotiations at the very beginning with your own version of the agreement, drafted by your patent attorney. Otherwise, the corporation takes the lead, and you'll be on the defensive.

Signing: Don't sign any agreement that can cause you to lose your patent rights if you default on any of the terms of the contract. In the event of default by either party, the remedy that is fair to both parties, and places you at the least disadvantage, is arbitration, not litigation.

Connecting:

The most effective approaches to marketing your invention are covered by five main strategies. These are arranged in the order of most effective first:

a. Produce a limited quantity of your invention, and sell it to strangers.

b. Offer a virtual product for sale, and show evidence that it will sell.

c. Offer a virtual product without evidence that it will sell.

d. Offer your invention and patent without a sales pitch.

e. Delegate the responsibility for licensing to an agent.

a) Produce a limited quantity of your invention, and sell it to strangers:

The objective of this approach is to obtain a sample of future possible sales in order to convince a licensing prospect that you truly have a marketable product. The main disadvantage to this approach is the cost to produce, which will usually be greater than your selling revenue. The advantages are that you will prove to yourself and your potential licensees (or your marketing channels, if you intend to produce) that your invention attracts buyers and is therefore capable of making a profit.

The cost to produce is determined by the manufacturing process, which, in turn, depends on tooling, materials, and time. Tooling is usually the most significant cost for small production runs. And nearly all manufacturing processes offer a spectrum of tooling options that affect the cost per unit produced. The simple rule is this: *The greater the investment in tooling, the lower the cost per unit.* For our purposes, that rule can be reversed:

The lower the investment in tooling, the higher the unit cost.

This is not an absolute rule. In some cases is doesn't hold because the purpose of a given process is the saving of time, not money. The term "rapid prototyping" may involve a relatively high tooling cost *and* also high unit prices for processes that produce limited quantities. But the objective is to get the parts almost overnight and beat the competition to the market.

An example of the rule in bold print above is that of a plastic part. Made on an automatic milling machine, the part may cost $3.00, and the tooling (essentially, a program) cost $150. This same part, injection molded, may cost less than a quarter, but the tooling—an injection mold—may cost $20,000 or more.

Obviously, for a market test, the safe approach is to pay the $3.00 for the part, and avoid the investment in tooling. (Yet, I've known inventors who spent more than $25,000 for a class A mold because they were unaware of any lower tooling-cost process.)

Therefore, *investigate the processes by which a given part can be made*. One of the surest ways to learn is to attend job-shop shows, and talk with vendors. Bring a print, or a prototype, and ask questions until the vendors stutter. Mainly, ask something like this: "What quantity range is usually economic for your process?" "If you were going to make only a hundred of this part, how would you do it?" "What is the process that is economic below the quantity range which is economic for you?" To find job-shop shows near you, just Google "job shop shows." (Be sure to use the quotes.) I just came up with more than 1,000 references. Another source is **www.jobshop.com**, which offers white papers through which you can learn about the various processes.

Whether licensing or testing the market before producing, it is prudent to preserve your capital. You'll need it for other expenses in developing your invention into a product.

Why sell it to strangers? Because selling it to friends and neighbors, or selling it in any way that offers a special advantage, proves nothing to you or anyone else. You've got to offer it on the Internet, or as a consigned item in a store, or to a catalog buyer, etc., in order to show that it will sell. Armed with unbiased sales data and a sell-sheet or a prospectus, your chances of reaching the negotiations stage are vastly improved.

Whether your objective is to license or establish distribution outlets for your product, the essential medium by which your communications will be made must include the sell-sheet, a.k.a. the brochure. I'll cover producing the sell-sheet further on.

b) Offer a virtual product for sale, along with evidence that it will sell:

The objective of this approach is to obtain an *estimate* of future possible sales, based on a survey rather than actual sales, in order to convince a licensing prospect that you *will* have a marketable product. This is generally the most practical approach for most inventors because the investment, although not inexpensive, is affordable. I maintain that if you can't afford this approach, you should abandon any further efforts.

Your object here is to solicit opinions from strangers (for instance, at a mall) on at least three basic questions:

- If you saw this product in a catalog, would you recognize its benefits?
- If you saw this product in a catalog, would you or your friends likely buy it?
- How much do you think most people would be willing to pay for it?

By virtual product I mean a product that exists only digitally in the memory of a computer, or on a CD or DVD. The simplest, of course, is the virtual photograph. Graphic artists who specialize in preparing 3-D drawings of concepts that don't physically exist, can create a very convincing "photo" of your not-yet-existing product, and this can be incorporated into a sell-sheet (brochure).

In some cases—especially those of products that lend themselves to demonstration—a DVD is more convincing than a static photo. The DVD can be produced with animation that demonstrates your product, but animation is very expensive. In most cases, you can save a few thousand dollars by having a physical prototype made, and taking a movie of someone demonstrating it. The movie should be taken using a digital video camera so that it can be transferred to DVD for sending out in quantity.

Your virtual prototype should be supported with convincing sales survey data if you've done your homework.

You can, of course, take actual photos of your physical prototype, if it is cosmetically acceptable. Even though you have at least one physical product at this point, it is still, in a sense, a virtual product because it doesn't exist for sale. Still another possibility for your sell-sheet is what I call a photo dummy. This is a physical model of your product that looks like your product eventually will look, but is not a functional prototype.

c) Offer a virtual product for sale *without* evidence that your product will sell:

Although offering a DVD virtual prototype or sell-sheet and letter alone is a weak strategy compared with offering these *along with* evidence of sales, if your budget won't allow you to do more, you might have some success with such an approach if you contact a relatively large number of prospects.

Write a letter explaining what your product is, and how it benefits the ultimate user. *Stress benefits* not features. You might tell something about the features after introducing benefits, but this should always be subordinate to benefits. (We inventors think in terms of features. Potential licensees or consumers want to hear *benefits*.)

d) Offer your invention and patent without a sales pitch:

Many inventors send off a copy of their patent, and a cover letter that makes no attempt to explain the benefits of the product that their invention will eventually become. This is a very weak approach, and generally doesn't get any results. Patents get reviewed by engineers and/or lawyers, and neither of these disciplines is marketing oriented. The only two corporate persons with whom a connection will pay off are the President and the Marketing Director (often a Vice President). In order to get through to these titles, you must first pass the legal requirement of signing away your rights other than those provided by your patent.

e) Delegate the responsibility for licensing to an agent:

Invention marketing agents that don't demand up-front money are difficult to find. They keep a low profile because they are otherwise besieged by inventors with ideas for an invention, rather than a true invention. The difference? An invention ready for licensing by an agent will have a patent, a sell-sheet, and, if possible, a "looks like, works like" prototype. Invention marketing agents don't have time to coach inventors, and get them up to speed. (That's what I do.) In fact, at least 80 percent of the inventions offered to them are rejected because licensing is difficult, and effective agents will only take on those inventions that offer high prospects for licensing. Ask what percentage of clients who contact them are taken on as clients. My advice is to avoid agents who take on more than 20 percent (one out of five). In my opinion, such agents are gamblers who will waste your time, and most likely not conclude a licensing deal for you.

You may find an agent advertising in *Inventors' Digest* (1-800-838-8808). Beware of those who want substantial up-front money for "market research" and "evaluations." A few hundred dollars may be fair compensation for their time, but when the price approaches a thousand dollars or more, be skeptical.

Seducing: (Preparing the sell-sheet—the essential piece of paper):

What is a sell sheet?

A sell sheet also known as a brochure, is usually a single sheet, 8-½ by 11 inches, with a color photo (or computer-simulated photo) plus black printing at least on the front side. It contains a picture or drawing of the product, a brief (one sentence) description of what it is and how it works, the benefits of the product to the ultimate user, and the product's features.

The sell sheet does not show the product's price, either wholesale or retail. Pricing should be handled by a separate price sheet. This avoids having to re-do the master if you change your pricing, or if you use it for different classes of trade (e.g. distributor, wholesaler, cataloger, retailer, individual.).

What purpose does a sell sheet serve?

In general, a sell sheet is a marketing tool that has several possible uses. For inventors it is used to convince catalog buyers or retail buyers that your "product" is worthy of being cataloged or sold in stores. I have used " " around the word *product* above because sometimes the sell sheet is prepared long before the product exists

The sell sheet can also be used to convince individual buyers to purchase your product. In this sense it can be a response document to cover a lot of ground, saving time preparing custom answers to inquiries. But because sell sheets, postage, and time are a significant expense, the kind of sell sheet that I am suggesting here is used mainly to *sell to the people who sell*.

A sell sheet should be accompanied by a tailored letter in nearly all cases, even if that letter is computer generated.

How do you make a sell sheet?

For many of us inventors the sell sheet demands writing and graphic skills beyond our capabilities. Making your own is not the equivalent of drilling and filling your own tooth. But the graphics and the copy writing must look professional if you hope to succeed. An amateurishly prepared sell sheet is perhaps worse than none at all.

To find services that prepare sell sheets look in your local yellow-page directory under any or all of the following categories: Graphic artists; graphic art studios; graphic printing services; graphic design; graphic designers; advertising agencies; artists; and printers. Sometimes the display ads for printers will include mention that they offer brochure and booklet preparation. Generally, these people are less expensive than those who advertise themselves as graphic artists. But be sure to see samples of their work.

Some of the above services are capable of creating a *virtual photograph* of your product, which is drawn, shaded, and textured on a computer, and exists only in the minds of the inventor and his or her artist, and in the memory of the computer. This isn't inexpensive, however. If you have a prototype that looks professional, it may be less expensive to have a professional photographer take pictures of it. Digital photography is acceptable, and may prove less expensive than film because the digital data can be used directly in the preparation of the sell sheet. Film photography requires scanning your photo, an extra process.

The biggest problem with preparing a sell sheet—even those prepared by some professionals—is that they are a hodgepodge of pictures, trite phrases, and dull sentences that praise features, rather than benefits. Even some of the professionals who prepare sell sheets have never read a book on how to write an ad, and a sell sheet is, above all, an ad. Not just any ad, but the single most important ad that your invention will ever depend on for success.

A model sell sheet: (Follow this plan for a rough draft of your sell sheet. Instruct the preparer that you want this model followed, and don't settle for anything but.)

- **At top of front page**: tag-line in bold print. A tag-line is a single brief sentence that expresses the major benefit to the ultimate user, (not, for example, the catalog or retail store buyer to whom you are pitching your product). You absolutely must attract attention and arouse interest within a few seconds with your tag-line. (Statements like "we are proud to announce" are a turn-off, and a waste of this precious space. No one gives a damn how proud you are except your parents or your spouse. Think only from the customer's viewpoint! A brief tag-line may have a sub-tag-line below it, like a book title often does. In general, avoid questions such as: *Do you spend too much time washing your car?* Questions are weak, and you don't always get the answer you want. *The new XYZ sponge washes your car in seven minutes flat!* And maybe a sub-tag in smaller print: *and saves an average of four gallons of hot water.*

- **Immediately below the tag-line, on the left,** place a color photo about 3 by 4 inches, (4 inches running vertically). Show your product being used by a person if appropriate. Get in close with the camera. Show hands only unless you need to show most of the person, such as washing a car, etc.

- **Immediately to the right of the photo** place a brief statement that defines your product. The XYZ sponge is . . .

- **Immediately below this brief statement** list, in bulleted form (like these statements) the benefits of your product. List in the order of strongest first.

- **Below the photo and the benefits** you might write about the features, but only if you relate how each feature creates one or more benefits. Don't talk about how you got the "great idea." No one except your mother is interested.

- **Half way across the bottom of the page, at the left,** place your company name and contact information, including your e-mail and your web site. To the right of this information place a blank rectangular box about an inch high. This box is to allow for rubber stamping other contact information, such as "Contact Jim Smith (your licensing agent), etc.

Negotiating and signing:

Inventors are usually lousy negotiators. Hire a professional from the Licensing Executive Society, (**www.les.org**) or any licensing agent who know what terms belong in a license, and how to compromise when necessary. Patent attorneys are usually good consultants, and sometimes even good negotiators. But most lawyers tend to be deal

You And Your Big Ideas

killers because they overly protect their clients to the point where no deal is good enough.

If you must do your own negotiations, always consult with a patent attorney on the terms of the agreement before signing. Once you sign, it's too late to start thinking about terms that are too advantageous to your licensee. As our old friend, Mark Twain has said: *It's easier to stay out than to get out.*

Richard Klar

RICHARD KLAR is... an intellectual property attorney with over twenty-five years of experience in patent, trademark and copyright law practicing in Mineola, NY. He is a member of the New York Bar and is licensed to practice before the U.S. Patent and Trademark Office. He has prepared and prosecuted US and foreign patent applications in various technologies including software, electrical, electronic, medical technologies and mechanical arts and business method patents. He has also secured copyright and US and foreign trademark registrations. He has litigated successfully in US District Courts throughout the countries including securing preliminary injunctions and summary judgments and has also successfully argued before the U.S. Court of Appeals for the Federal Circuit cf. *American Permahedge v. Barcana*, 105 F.3d 1441, 41 U.S.P.Q.2d (BNA) 1614 (Fed. Cir. 1997)

Richard has considerable experience in intellectual property transactional matters, including reviewing IP portfolios for mergers and acquisitions and licensing and contractual matters involving intellectual property rights. He has lectured at bar associations and trade show venues over the years on intellectual property law.

Contact:

Richard B. Klar, Counselor at Law
145 Willis Avenue, Mineola, NY 11501

Phone: (516) 248-8800

Website: www.nypat.com   Email: richardklar@msn.com

Author THE UNITED STATES SUPREME COURT'S DECISION IN EBAY V. MERCEXCHANGE: HOW IRREPARABLE THE INJURY TO PATENT INJUNCTIONS? (forthcoming Summer 2008 –Widener Law Review Vol14, No 1)

Author EBAY INC. V MERCEXCHANGE , L.L.C.: THE RIGHT TO EXCLUDE UNDER U.S. PATENT LAW AND THE PUBLIC INTEREST

27 Whittier L.Rev. 985-995 (2006) Reprinted in October 2006 issue of the Journal of the Patent and Trademark Office Society Vol. 88, No. 10

THE RESOURCE GUIDE

D o your research
A good way to find out about your particular sector and industry is to do a simple internet search by typing "trade magazine for"…

Most of the time subscriptions are free and reading these publications will help you familiarize yourself with the language, understand the industry standards and recent innovations. You don't have to become an expert, but you have to be able to talk to people using the language they understand. It can also help you make valuable contacts with people whose specific knowledge or expertise can help you. You never know who you're going to run into or where you will find them.

Research also helps you gain confidence when you pick up the phone to speak to prototype designers and engineers, manufacturers and licensees, patent attorneys and agents. In other words, the more you know, the more success you can achieve in less time. You can do it all yourself or you can leverage the information, experience and reputations of the people you network with.

SUPPORT GROUPS

EVERYDAY EDISON
www.everydayedison.com

INVENTORS & ENTREPRENEURS CLUB
This is the group I organized in Suffolk County, New York.
www.iesuffolk.com
email: info@iesuffolk.com

UNITED INVENTORS ASSOCIATION
This is a not-for-profit organization that helps inventor groups and inventors in North America. Much information and an excellent collection of books for inventors can be found in its Inventor's Bookstore.
www.Uiausa.org

FUNDING SOURCES
www.en.wikipedia.org/wiki/Venture_capital
www.ezinfofind.com/download.asp

GOVERNMENT AGENCIES

SMALL BUSINESS DEVELOPMENT

Help with starting a small business and writing business plans:

www.sba.gov

www.sbdc.gov

GLOBAL LICENSING INFORMATION

www.licensing.org

US PATENT & TRADEMARK OFFICE

File patents and trademarks, conduct searches, find patent agents and attorneys and more:

www.uspto.gov

INTERNET PROVIDERS

ALTERNATE PATENT SEARCH LOOKUP TOOL
www.google.com/patents

DOMAIN NAMES
To investigate domain names…

www.whois.net

www.godaddy.com

www.register.com

SEARCH ENGINGES
www.google.com/

www.yahoo.com

www.msn.com

www.ask.com

WEBSITES
For basic site development and hosting:

www.homestead.com

www.godaddy.com

SERVICES

AGREEMENTS
NON-DISCLOSURE AGREEMENT
www.score.org
www.asktheinventors.com

BUSINESS PHONE SERVICE
www.att.com

JOB SHOP SHOWS
www.jobshop.com

LICENSING AGENTS
LIMA – Licensing Industry Merchandisers' Association
www.licensing.org

Periodical Resource for Agents:
www.licensemag.com

LICENSING EXECUTIVE SOCIETY
www.les.org

MANUFACTURER'S LISTINGS
www.Thomasnet.com

MARKETING AND DESIGN SERVICES
www.vertigonewyork.com

PACKAGING
www.packagingdiva.com

PATENT SEARCH
Patent Search International
www.patentsearchinternational.com
I've dealt with Ron Brown.

PRODUCT DEMO VIDEOS
www.nationalmediaconnection.com

PROFESSIONAL ADVICE
www.score.org
www.asktheinventors.com

SUPPORT PROVIDERS
www.Elance.com
www.Coroflot.com
www.Guru.com

TRADE SHOW SCHEDULES
www.tradeshowweek.com
www.tsnn.com
www.yankeeinventionexpo.org

PUBLICATIONS

Subscriptions:

ENTREPRENEUR MAGAZINE
Has the largest newsstand circulation of any business monthly (500,000)
www.entrepreneur.com
1.800.274.6229

INC. MAGAZINE
Provides small business resources
www.inc.com

INVENTORS DIGEST MAGAZINE
The only magazine dedicated exclusively to inventors.
www.inventorsdigest.com
Subscriptions/Customer Service
Internet/Print advertising information
1.800.838.8808

Authors:

DON DEBELAK
Don has written 12 books and provides free tips to innovators and inventors.
www.DonDebelak.com

JACK LANDER
Jack is a successful inventor and the author of books and condensed reports for inventors.
www.inventor-mentor.com

BOB PROCTOR
Author, lecturer, business consultant, entrepreneur.
www.bobproctor.com

Source Books for your Reference Library:

12 Amazing Secrets of Millionaire Inventors: Smart, Simple Steps for Turning Your Brilliant Product Idea into a Money-Making Machine by Harvey Reese Publisher: Wiley (August 31, 2007)

The Inventor's Pathfinder: A Practical Guide to Successful Inventing by James L Cairns Publisher: iUniverse, Inc. (Sept 29, 2006)

What Every Inventor Needs To Know About Business & Taxes by Stephen Fishman Publisher: NOLO; 2 Pap/Cdr edition (May 10, 2005)

Product Idea to Product Success: A Complete Step-by-Step Guide to Making Money from Your Idea by Matthew Yubas Publisher: Broadword Publishing (Jan 2004)

The Inventor's Bible: How to Market and License Your Brilliant Ideas by Ronald Louis Docie Publisher: Ten Speed Press; Rev Sub edition (March 2004)
Secrets from an Inventor's Notebook by Maurice Kanbar Publisher: Penguin (Non-Classics); 1st edition (February 15, 2002)

What the World Needs Now: A Resource Book for Daydreamers, Frustrated Inventors, Cranks, Efficiency Experts, Utopians, Gadgeteers, Tinkerers and Just About Everybody Else by Stephen Johnson Publisher: Ten Speed Press U.S.; Rev Ed edition (Mar 2001)

The Inventor's Kit: A Complete Workbook for Filing U.S. Patents, Trademarks & Copyrights by W. J. Scott Murphy
Publisher: Symposium Publishing; New Rev edition (May 31, 2001)

The Complete Idiot's Guide to Cashing in On Your Inventions by Richard Levy
Publisher: Alpha (September 28, 2001)

The Inventor's Notebook
by Fred E. Grissom, David Pressman
Publisher: NOLO (November 2000)

Stand Alone, Inventor!
by Robert G. Merrick
Publisher: Lee Publishing (October 1997)

Printed in the USA
CPSIA information can be obtained
at www.ICGtesting.com
CBHW061546121124
17304CB00034B/346